Programming the Photon

Programming the Photon

Getting Started with the Internet of Things

Christopher Rush

New York Chicago San Francisco
Athens London Madrid
Mexico City Milan New Delhi
Singapore Sydney Toronto

Library of Congress Control Number: 2016933233

McGraw-Hill Education books are available at special quantity discounts to use as premiums and sales promotions or for use in corporate training programs. To contact a representative, please visit the Contact Us page at www.mheducation.com.

Programming the Photon: Getting Started with the Internet of Things

1 2 3 4 5 6 7 8 9 0 DOC/DOC 1 2 1 0 9 8 7 6

ISBN 978-0-07-184706-3
MHID 0-07-184706-5

This book is printed on acid-free paper.

Sponsoring Editor
Michael McCabe

Editorial Supervisor
Stephen M. Smith

Production Supervisor
Lynn M. Messina

Acquisitions Coordinator
Lauren Rogers

Project Manager
Hardik Popli

Copy Editor
Lisa McCoy

Proofreader
Rajni Negi,
Cenveo Publisher Services

Indexer
Jack Lewis

Art Director, Cover
Jeff Weeks

Illustration
Cenveo Publisher Services

Composition
Cenveo Publisher Services

About the Author

Christopher Rush has a degree in computer science and has spent the last 10 years working for an electronics distribution company as a product manager for single-board computing. He also runs a MakerSpace blog (www.rushmakes.com) providing reviews, tutorials, and user guides for popular development boards and accessories, including Raspberry Pi, Arduino, BeagleBone, and others. Mr. Rush is the author of *30 BeagleBone Black Projects for the Evil Genius,* also published by McGraw-Hill Education.

CONTENTS AT A GLANCE

CONTENTS

PREFACE

This book is the perfect introduction to programming the Particle Photon development board. The Particle Photon is a true Internet of Things device that lets you write code and create electronic projects using the cloud. It is fully capable of acting as the brains of your projects while expanding their capabilities by using the Internet to remotely control and collect data.

Luckily, the Photon platform has adopted the Arduino-style programming language while also introducing its own programming features. This opens you up to the vast amount of resources from the world of Arduino that are available to you, including existing projects and examples.

Why the Photon? The Photon board was developed by the team at Particle and was introduced to the world in November 2014, priced at only $19. It is one-of-a-kind, offering a unique hardware and software experience to you using the Particle cloud, which can be programmed through the Web IDE. The Photon board supersedes the Particle Core, which was funded through a Kickstarter campaign that raised over half-a-million U.S. dollars, and it comes equipped with the Broadcom BCM43362 Wi-Fi chip rather than the TI CC3000.

The purpose of this book is to get you started with creating your own hardware projects with the Particle Photon. You do not need any previous experience wiring circuits or programming, but a general use of computer skills would be highly advantageous. *Programming the Photon* is written to give you a wide variety of experiences and a basic understanding of the many capabilities of the Photon board. This book covers only the basics of how to program the board, on the assumption that you will then expand those skills on your own for your future projects.

I would love to hear your thoughts on this book and would encourage you to contact me through www.rushmakes.com or Twitter (https://twitter.com/chrisrush85).

Christopher Rush

ACKNOWLEDGMENTS

I would like to thank Mike McCabe and the team at McGraw-Hill Education, who have been very supportive and a pleasure to work with once again.

I would also like to dedicate this book to my partner Jennifer Wozniak, who, as always, gives me encouragement and motivation throughout; I would be lost without her by my side.

Programming the Photon

1

Introduction to the Photon

In this chapter you will learn a bit about microcontrollers such as the Arduino as well as the Internet of Things (IoT). The Photon board is a new development board based on its predecessor, the Core, with some new hardware and software features that make it far superior. We will take a look at all those features and compare both boards.

Microcontrollers

A microcontroller is essentially a computer that can control multiple inputs and outputs using some form of programming language. Microcontrollers come in all sorts of shapes and sizes, with the most popular platform being the Arduino. Arduino boards provide a low-cost, easy-to-use technology to create small electronic projects. Modern conventional microcontrollers can be connected to a computer using universal serial bus (USB) to power the board as well as to program the microcontroller; however, they can also easily be removed from USB once the program has been uploaded and powered using some sort of portable battery device and work independently.

Other popular microcontrollers are the Raspberry Pi and BeagleBone boards. Both of these boards are more advanced than the standard Arduino boards and have some video output for connecting to a visual display with a basic operating system such as Debian. These boards feature a vast array of hardware abilities, which can take your electronic projects to another level by providing more storage, input/output pins, faster processing, and audio/video output. All of these options are great, but when you need to connect your projects to the Web, you almost certainly need extra hardware, such as shields or USB dongles, and this adds considerable cost to your projects—sometimes more than the microcontroller boards cost themselves. Some variations of boards do include built-in Wi-Fi or Bluetooth technology, such as the Arduino Yun, but this board is still rather expensive at more than $70; once you have added your electronic hardware and various other costs, then your project can easily be in excess of $100.

So, What Is the Photon?

The Particle Photon is a single microcontroller development board much like the Arduino Nano, with a small form factor but with the added feature of having a built-in Wi-Fi module that you can control and program over the Internet using the Particle cloud. Once connected to your local Wi-Fi network, you can also control and program your Photon with your smart phone using the Particle app in either IOS or Android operating systems. The Photon board itself has lots of pin headers down each side of the board; these pins act as the inputs and outputs for the microcontroller. These general-purpose pins can be connected to sensors or buttons to listen to the world, or they can be connected to lights and buzzers to put on a show. There are also pins that allow you to power your Photon board, motors, and outputs of your device. In addition, the Photon board comes with some built-in hardware features such as buttons and light-emitting diodes (LEDs), which make things a lot easier when configuring the Photon board:

- The SETUP button is on the left, and the RESET button is on the right. You can use these buttons to set your device's mode.
- The RGB LED is in the center of your Photon just above the module. The color of the RGB LED tells you what mode your Photon board is currently in.
- The D7 LED is next to digital pin 7 on your Photon. This digital pin will turn on the LED when pin 7 is set to HIGH.

Particle Photon versus Spark Core

The Photon board is the predecessor of the Core, both of which are produced by Particle. When compared side by side, both boards look similar, and it is difficult to tell them apart. The main difference is in the hardware aspect, as the Photon board uses a different Wi-Fi chip than the Core and also has a faster processor and more random access memory (RAM).

The pins on both boards are almost the same; hence most of the experiments in this book will work well with the Core too. There are a few changes to the Photon that add a lot of benefit, such as digital-to-analog convertor (DAC) and wakeup (WKP) pins instead of having A6 and A7 on the Core.

The Internet of Things

The Internet of Things (IoT) is media terminology for taking what are considered dumb electronic devices and connecting them to the Internet. Once connected to the Internet, you can control these devices through your Web browser, which will send Hypertext Transfer Protocol (HTTP) requests to a Web server and send back the information displayed. You can connect a whole range of devices and sensors for applications such as

- Home automation
- Weather stations
- Robotics
- Air pollution monitoring
- Environmental sensing
- Smart logistics
- Location tracking
- Health monitoring

More and more IoT devices are becoming available on the market these days, such as smart thermostats or Philips Hue lamps, which allow the user to control home heating aspects or mood lighting. With this big boom in the IoT, makers and hobbyists have become more intrigued with creating their own smart projects, and the Photon board allows them to do just this while keep the cost down to a mere $19—one of the least expensive boards on the market today.

Because so many makers and hobbyists are creating new IoT projects, it makes sense to create a simple framework for both the hardware and software that provides a simple, easy-to-use system for any level of skill. This is why the Particle team has created such a system based on the popular

Arduino software, transforming what could be quite complex technology into an open-source product that is easy to use for every user.

Particle Cloud

The hardware part of the framework is the Photon board itself, which is based on the popular Core module funded through the crowd-funding website Kickstarter. The Photon board has been designed to be backward compatible with the Core, and as such, most of the experiments in this book will also work with the Core.

Particle has created a software framework for the hardware that allows users to interact their hardware with other technology and devices over the Internet, and the two elements work hand in hand easily. An IoT device that uses the Photon might turn on a consumer device using a relay or similar circuitry. In this case a user might access a webpage or a mobile app that has a button to switch the device on or off. When the user clicks the button on the webpage, it then sends a message or string of data to the Particle cloud service, which then forwards that message on to the Photon board and switches on the device. If the Photon board had some sensors connected to it, then the cloud system would simply work in the reverse order, where instead of the Web service sending information to the cloud when you clicked the button, the Photon board would send the sensor information to the cloud and then to the Web server to be displayed on the Web. The whole Particle framework makes this work seamlessly without being too complicated for end users—all you have to do is register your Particle Photon board with your Particle cloud account and you are ready to go.

The Photon Board

The Photon board itself is a very small form factor, as you can see in Figure 1.1.

The two buttons on the Photon board, SETUP and RESET, allow you to configure your Wi-Fi credentials and restart the device if needed. Combined together they also trigger a full factory reset just in case you get into a bit of trouble with your device.

At the top of the board you will find the micro USB port, which is used to power the board, as well as connecting it to your computer for USB programming if required.

Figure 1.1 *Photon board.*

The Photon board has a built-in chip antenna, which is fine for most indoor applications, but the Photon also features an external socket for connecting a Wi-Fi antenna for range extendibility and a directional antenna. The Photon board is configured by default to always choose the most reliable method if both the chip antenna and external antenna are used. The antenna can also be manually selected in the firmware.

Summary

By now you should be itching to get started. The Photon board is a great device for creating IoT projects in the maker community, as well as for developing commercial consumer products. The next chapter will show you how to set up your Photon board and start programming your first project.

2

Getting Connected

In this chapter we will show you the many ways of connecting your Photon board to the Particle cloud and start programming. By far the easiest way to connect to your Photon is using the Tinker application on your smart phone for either iOS or Android, but just in case that doesn't work out for you or you do not have a smart device, we will show you other methods of connecting to your Photon.

Board Features

Before we get started, it's important that you fully understand the board itself and some of the useful features, including the most important RGB (red, green, blue) light-emitting diode (LED), which will become your ultimate tool when understanding what the board is doing.

Looking at Figure 2.1 of the Photon board, you can see there are lots of pin headers, a few buttons, and a bright flashy LED. There are two buttons located on the Photon: the RESET button, which is located on the right when holding the Photon with the universal serial bus (USB) port at the top, and the SETUP button, which is located on the left side of the Photon.

Holding down the RESET button will effectively initiate a hard reset of the device, powering the device down and then repowering the device back up. If you have loaded a program into the Photon and you are experiencing some issue, then this is a good way to restart the program.

The SETUP button does a number of things when held down. When held down for 3 seconds, this puts the Photon into something called *Smart Config*—this allows your Photon to connect to your local Wi-Fi network and will be

Figure 2.1 *Photon board.*

indicated by a blue flashing LED. When the SETUP button is held down for up to 10 seconds, this clears the Photon's Wi-Fi memory and deletes any saved Wi-Fi credentials. This is particularly useful if you want to connect to another Wi-Fi network or are experiencing issues with the current network. If you hold down the SETUP button and then instantly tap the RESET button, this initiates the *Bootloader* mode after 3 seconds. The Bootloader mode enables you to reprogram the Photon through USB or using the Joint Test Action Group (JTAG) add-on board. The Bootloader mode will be indicated by a flashing yellow LED. If you have entered this state by mistake, you can simply hit the RESET button again to exit it. If your Photon board is not responding after a hard reset, then the last option is a factory reset of the board itself, which will wipe everything off and restart your Photon as if it was newly out of the box again. You can do this by holding down both the SETUP and RESET buttons for 10 seconds—the LED should flash white quickly and then turn another color when it has been reset.

On the Photon board you should be able to see two LEDs. There is one RGB LED in the middle of the board, which shows you the status of your Photon's network connection. There is also one other small blue Surface Mount Diode (SMD) LED, which is the user LED; this is connected to pin D7, so when you turn the D7 pin HIGH or LOW, it turns the blue LED either ON or OFF. The RGB LED shows the following states:

- **Flashing blue** *Listening* mode; waiting for network information.
- **Solid blue** *Smart Config* complete; network information has been found.

- **Flashing green** Connecting to the local Wi-Fi using the credentials found.
- **Flashing cyan** Connecting to the Particle cloud.
- **Quick flashing cyan** Initiating a handshake with the Particle cloud.
- **Slow pulsating cyan** Successfully connected to the Particle cloud.
- **Flashing yellow** *Bootloader* mode; awaiting for new code via USB or using the JTAG shield.
- **White pulsing** Startup; the Photon is powered on.
- **Flashing white** *Factory reset* has been initiated.
- **Solid white** *Factory reset* has completed and the Photon is resetting.
- **Flashing magenta** The Photon is updating the firmware.
- **Solid magenta** The Photon may have lost connection to the Particle cloud; by pressing RESET the Photon board will attempt to update the firmware once more.

Although these LED statuses give us a good, clear indication as to what the Photon board is doing, it is also useful when things don't go so well. The RGB also flashes errors with a red LED flash. These errors may include the following:

- **Two red flashes** Connection failure due to a bad Internet connection.
- **Three red flashes** There is an Internet connection but the Particle cloud is inaccessible. Visit the Particle website for an up-to-date status of the cloud.
- **Flashing yellow/red** The Wi-Fi credentials provided to the Photon board are incorrect and as such cannot connect to the local Wi-Fi.

The Photon has 24 header pins that are clearly visible from underneath; each pin is also clearly labeled on the silk screen (top). These pins are clearly labeled with the following:

- **VIN** Indicated as voltage input; connect an unregulated power source to this pin with a voltage between 3.6 V and 6 V (max) to power the Photon.

NOTE *When powering the Photon through the USB port, the VIN pin should not be used.*

- **3V3** This pin, as the name suggests, gives a regulated output voltage of 3.3 V that can be used to power your circuits. This pin can also be used to power the Photon if you have a regulated 3.3-V power source.

NOTE *Powering the Photon using the 3.3-V rails is not recommended. Any overvoltage may cause permanent damage to the board.*

- **VBAT** Supply to the internal Real Time Clock (RTC), back registers, and Static Random Access Memory (SRAM) when the 3V3 is not present (1.65 to 3.6 V).
- **RST** This pin allows you to reset your Photon when you connect it to one of the ground pins on the Photon board.
- **GND** The GND, or ground, pins are used to connect any positive voltage to ground.
- **D0** to **D7** These pins represent a digital input/output value from an electronic circuit or device. These pins are essentially your world. They cannot read any analog inputs/outputs from components such as sensors. Some of the digital pins also have additional features, such as support for peripherals such as Serial Peripheral Interface (SPI) or JTAG.
- **A0** to **A5** In contrast to the digital pins, there are an additional six GPIO pins; these pins are the same as D0 to D7 except they are analog pins, which means they can read values from analog sensors.
- **Tx** and **Rx** These pins are used for communication over serial/ Universal Asynchronous Receiver Transmitter (UART). Tx represents the transmitting pin, and Rx represents the receiving pin in the serial communication.
- **WKP** Active-high wakeup pin; wakes the module from sleep/ standby modes. When not in use, the WAKEUP pin can be used as a GPIO, analog-to-digital convertor (ADC), or pulse width modulation (PWM) pin.
- **DAC** Twelve-bit digital-to-analog output and also a digital GPIO. DAC is used as DAC or DAC1 in software, and pin A3 is used as a second DAC output as DAC2.

In addition to the GPIO pins, some of the analog and digital pins can be used as PWM pins using the function analogWrite(). These PWM pins do something called pulse width modulation, and they give the effect of increasing or decreasing the timings of switching something on or off—for example, you can dim the brightness of an LED or speed up a motor. The Photon has up to nine PWM pins: D0 to D3, A4, A5, WKP, Rx, and Tx.

Getting Connected

To power up your Photon board, simply plug the included USB cable into the Photon, as shown in Figure 2.2, and the other USB end into your computer/laptop or USB power supply. Once powered up, your Photon board should start blinking blue. If you intend to use the Photon with a u.FL connector, make sure the antenna is connected properly. If this is your first time, you should see the blinking blue LED. If you see another color LED, hold down the MODE button until the LED starts to blink blue and then continue.

Connecting to Mobile Smart Device

If you are going to set up your Photon board using your smart phone through either an iOS or Android device, you can locate the application in the app store by searching for "Particle" and download it free of charge.

When you first launch the application on your device, you will see the login screen as shown in Figure 2.3. Every Photon device must be registered on the Particle cloud using its unique identification number at the point of manufacture. If you have already signed up to the cloud, go ahead and log in; otherwise, register with your details for an account on the Particle cloud—it should only take a few minutes.

Figure 2.2 *Photon board connected to a laptop through the USB cable.*

Figure 2.3 *Particle application login screen.*

Once you have logged in, you need to make sure your smart device is connected to the Wi-Fi network that you wish to connect your Photon board to; otherwise, you will not be able to communicate and program it to do lots of amazing things. Your Wi-Fi network name should appear on the next screen in the SSID box. All you need to do now is enter your Wi-Fi password and click Connect and let the magic happen. It may take a few moments to connect, so be patient. The Photon board should go through the following colors in this order:

- **Blinking blue** Listening for Wi-Fi credentials
- **Solid blue** Getting Wi-Fi information from the Tinker application
- **Blinking green** Trying to connect to the Wi-Fi network
- **Blinking cyan** Establishing a connection to the Particle cloud
- **Blinking magenta** Updating to the newest firmware
- **Pulsating cyan** Successfully connected

Once successfully connected, you should see the screen in Figure 2.4, which shows you all the GPIO pins available for programming.

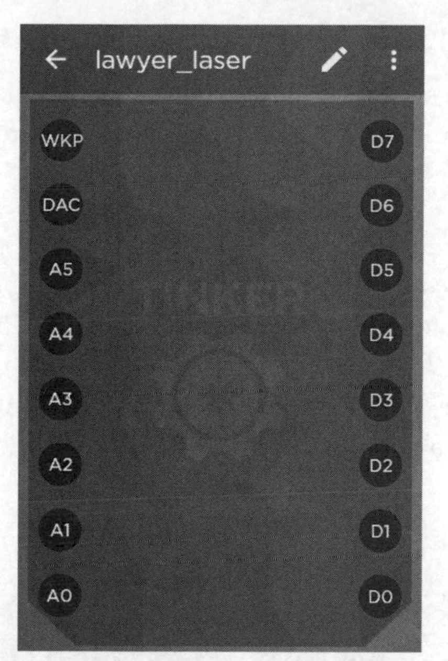

Figure 2.4 *Tinker GPIO pin configuration.*

Troubleshooting

From time to time your Photon board may not work in the way you want it to; this isn't a problem, as a few tweaks here and there will get it working in no time. Here are some guidelines to help you on your way:

- If your smart device has not detected any Photon devices, make sure you are connected to the same Wi-Fi network. If the Photon board is blinking blue, try giving it another go.
- If your Photon board is blinking green but not getting cyan, make sure you have entered the correct Wi-Fi credentials by trying again. Hold down the SETUP button until it is blinking blue and try entering your details again.
- If your Photon is pulsating cyan but the Tinker app did not find any devices, this means that the particular Photon board has not yet been registered to an account.

Connecting over USB

If you do not have a smart mobile device, you can connect the Photon board to your Wi-Fi network over USB by communicating through serial.

NOTE *This will work only if the Photon board is in Listening mode.*

The first thing we need to do is download a serial terminal application that we can use to communicate with the Photon. If you are using a Windows operating system, I highly recommend using PuTTY; you will also need to install the Windows driver for the Photon (https://s3.amazonaws.com/spark-website/Spark.zip). If you are using an Apple Mac computer, CoolTerm provides a good graphic user interface (GUI) and is user friendly (see Figure 2.5).

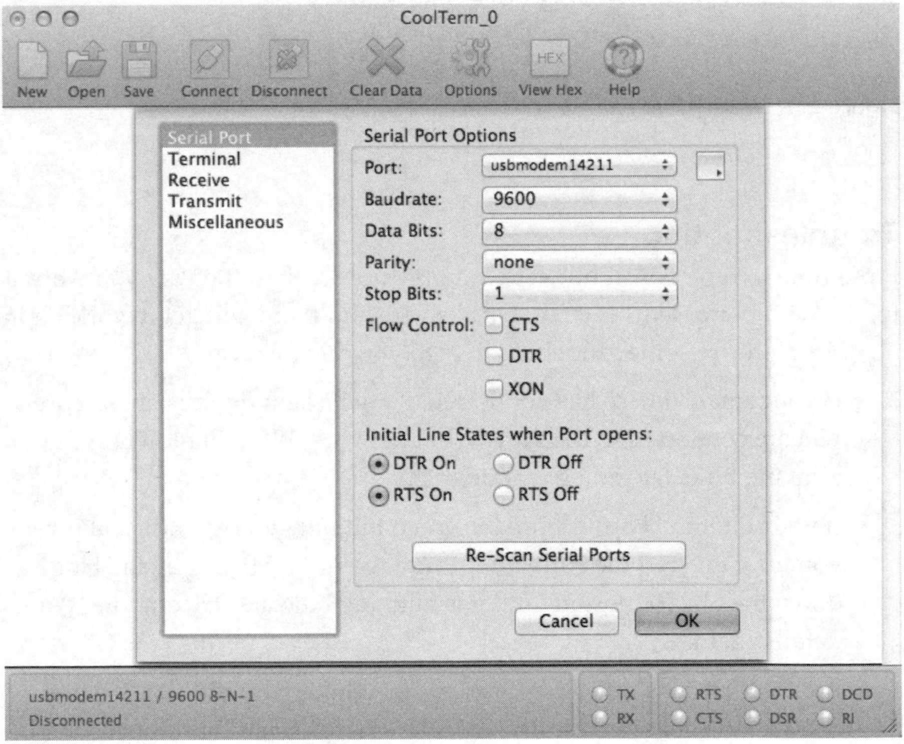

CoolTerm_0
New Open Save Connect Disconnect Clear Data Options View Hex Help
Serial Port
Terminal
Receive
Transmit
Miscellaneous
Serial Port Options
Port: usbmodem14211
Baudrate: 9600
Data Bits: 8
Parity: none
Stop Bits: 1
Flow Control: CTS
DTR
XON
Initial Line States when Port opens:
DTR On DTR Off
RTS On RTS Off
Re-Scan Serial Ports
Cancel OK
usbmodem14211 / 9600 8-N-1
Disconnected
TX RX RTS CTS DTR DSR DCD RI

Figure 2.5 *CoolTerm for Mac OS.*

When you have downloaded and installed the software, plug your Photon board into your computer via USB. When the Photon is in Listening mode (blue flashing light), open up a serial port for USB in the applications settings, which should be as follows:

- Baud rate: 9600
- Data bits: 8
- Parity: none
- Stop bits: 1

Once you have opened up a serial connection, you will have two different commands available by hitting either W or I on the keyboard, and here is what they do:

- **W** Set up your Wi-Fi Service Set Identifier (SSID) and password
- **I** Reads out the Photon board's unique ID

NOTE *If you connect your Photon board over USB for the first time, you will also be required to manually claim your Photon to connect to your account.*

Manually Claiming Your Photon

Once your Photon is connected to your Wi-Fi network, it also needs to be claimed to connect it to your account. This lets you control your Photon board and keeps anyone else from doing so. If you use the Particle mobile application, your Photon is automatically claimed to your account; however, if you connect your Photon over USB or if the claiming process was unsuccessful, then you need to claim it manually by using the following steps.

The easiest way to manually claim your Photon is to connect your Photon through USB serial and request the Photon's ID using the "I" command; you can then claim it via the Particle Build website. You device ID should be displayed as follows:

```
# Photon ID
55ff6806498949532909 2587
```

Once you have this unique Photon ID, open up the Particle Build page (http://www.particle.io/build) and click the Devices icon. Click the button that says Add New Device and enter your Photon's ID in the text box as shown in Figure 2.6.

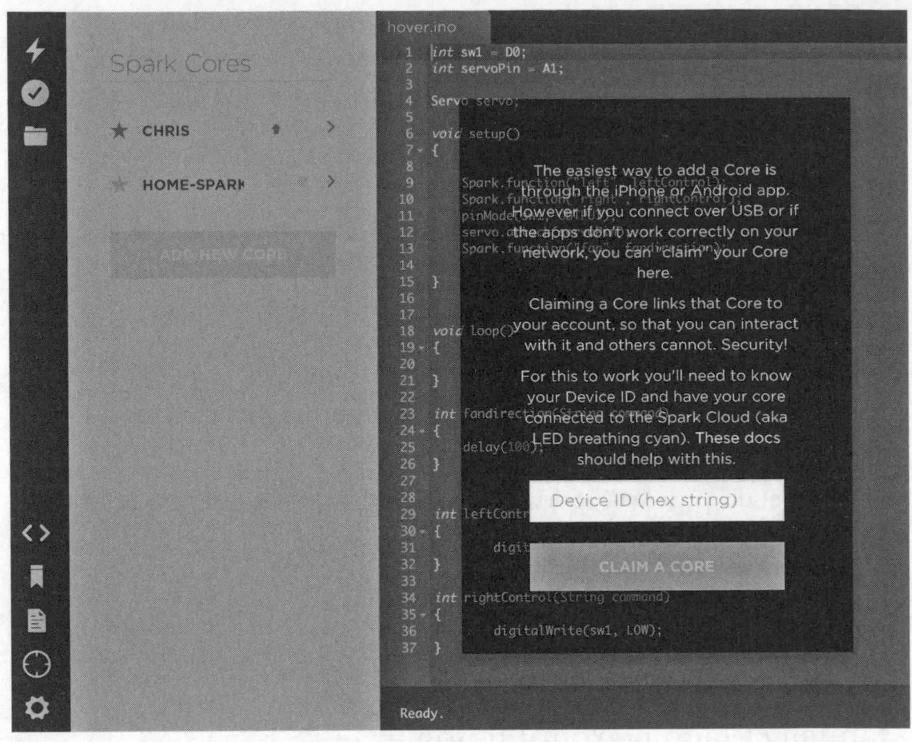

Figure 2.6 *Add a Photon through the Particle Build website.*

Using Tinker

After connecting your Photon to the Wi-Fi network using the app, you will be presented with the Tinker section, which makes it easy to start playing around with the GPIO pins on your device without actually having to do any programming. This is perfect for early project development and just playing around with things.

In Figure 2.4 you can see that the app displays 16 pins in vertical rows, similar to how the GPIO pins are set out on the board itself. To begin click on any of the pins—you should see a small menu pop up showing all the available functions for that particular pin. Each pin has up to four different functions:

- **digitalWrite** Sets the pin to either HIGH or LOW, which either connects it to the 3.3-V power rail or to the GND (ground). Pin D7 is connected to an on-board LED. If we use this pin as an example, when you change the pin D7 to HIGH, the LED will turn on and when you set the pin to LOW, it will turn off.

- **analogWrite** This function sets the pin to a value between 0 and 255, where 0 is the same as setting a digital pin to LOW and 255 is the same as setting a digital pin to HIGH. This sort of function sends a voltage between 0 and 3.3 V. Because this is a digital board, it uses something called PWM to emulate an analog signal. A good example would be to dim an LED by adjusting the analog value.
- **digitalRead** This will read the digital value of a pin, which, as we know, can either be HIGH or LOW. If you connect the pin directly to the 3.3-V pin, then the value read will be HIGH. If you connect it to GND, then the value read will be LOW. If the value read was anywhere in between these values, then it will give a reading of whichever one it is closest to; this is not advisable, as the readings could never be accurate.
- **analogRead** This will read any analog value from the analog pins on the board. The values read can be between 0 and 4095, where 0 is LOW and 4095 is HIGH (3.3 V). These analog pins are from A0 to A5. Typically these pins are used to read sensor values such as light and temperature.

Now that we know what functions we have for each pin, we can easily change the function of the pin simply by tapping and holding on the pin number, and the function select menu will be shown. Have a go and play around some more, flash some LEDs, read the temperature, drive some motors, or turn some servos.

When you first receive your Photon device, the firmware loaded onto the board is the default application for the Tinker application. If you have your own firmware loaded and want to use the Tinker app again, you can simply revert to a factory reset and the default firmware will load on to the board once again.

The easiest way to reflash your Photon is actually using the Tinker mobile application, using either iOS or Android devices:

- **iOS** Tap the list button on the upper-left side, then tap the arrow next to your device, and then tap the Reflash Tinker button that appears in the pop-out menu.
- **Android** With your selected Photon, tap the options button that appears in the upper-right side and then tap the Reflash Tinker option in the drop-down box.

Tinker API

When the Tinker firmware is installed on your Photon device, it responds to certain application programming interface (API) requests from the Tinker app on your mobile device, which in turn replicates the four functions of the GPIO pins. These API requests not only are read using the Tinker app, but you also can make requests from other applications, so you can quite easily create your very own Web or mobile application based around the firmware. The following section will guide you through making some simple requests based on the four GPIO functions available.

DigitalWrite sets the pin to either HIGH or LOW, which in turn connects to 3.3 V or to GND. Previously we know that pin D7 is connected to an on-board LED on the Photon board. If we set this pin to HIGH, the LED turns on, and when we set it to LOW, it turns off. The following code is the API request sent to the Photon to accomplish this task:

```
POST /v1/devices/{DEVICE_ID}/digitalwrite

# EXAMPLE REQUEST IN TERMINAL
# Core ID is 0123456789abcdef
# Your access token is 123412341234
curl https://api.spark.io/v1/devices/0123456789abcdef/digitalwrite \
-d access_token=123412341234 -d params=D0,HIGH
```

The parameters must be a pin number followed by the value, which is either HIGH or LOW. If the request has succeeded, it will return a value of 1 and –1 if it fails.

AnalogWrite sets the pin value in the range from 0 to 255, where 0 is the lowest value (GND) and 255 is the highest value (3.3 V). As previously mentioned, we are using a digital system, so it would not be possible to create an analog signal, but what we can do is emulate an analog signal using something called pulse width modulation or PWM for short. A good example of using PWM, and probably the easiest when starting out, is to use the analogWrite function to dim an LED. The following code is used to send the API request to the Photon:

```
POST /v1/devices/{DEVICE_ID}/analogwrite

# EXAMPLE REQUEST IN TERMINAL
# Core ID is 0123456789abcdef
# Your access token is 123412341234
curl https://api.spark.io/v1/devices/0123456789abcdef/analogwrite \
-d access_token=123412341234 -d params=A0,215
```

The parameters here must be the pin number followed by the integer value ranging from 0 to 255. As before, the return value will be 1 for success and –1 if it fails.

DigitalRead will read the value from one of the digital pins on the board; this value can be either HIGH or LOW. This API request is as follows:

```
POST /v1/devices/{DEVICE_ID}/digitalread

# EXAMPLE REQUEST IN TERMINAL
# Core ID is 0123456789abcdef
# Your access token is 123412341234
curl https://api.spark.io/v1/devices/0123456789abcdef/digitalread \
-d access_token=123412341234 -d params=D0
```

The parameter set must be the pin number—either A0 to A5 or D0 to D7; the return value will either be a 1 or –1.

AnalogRead will read an analog value of the pins labeled A0 to A5, which can be a value between 0 and 4095 where 0 is LOW and 4095 is HIGH. Only the analog pins can handle these values. Typically analog pins are used to read values from different types of sensors. The API request is as follows:

```
POST /v1/devices/{DEVICE_ID}/analogread

# EXAMPLE REQUEST IN TERMINAL
# Core ID is 0123456789abcdef
# Your access token is 123412341234
curl https://api.spark.io/v1/devices/0123456789abcdef/analogread \
-d access_token=123412341234 -d params=A0
```

The return value will be between 0 and 4095 if successful and –1 if it fails to read the value.

Running Tinker Alongside Your Scripts

If you have already played around with running your own firmware programs on the Photon, then you soon realized that you cannot use the Tinker app while your code is running. Well, now you can. Combine the following code with your firmware and flash it to your Photon and you will be able to run your program while tinkering away:

```
int tinkerDigitalRead(String pin);
int tinkerDigitalWrite(String command);
int tinkerAnalogRead(String pin);
int tinkerAnalogWrite(String command);

//PUT YOUR VARIABLES HERE
```

```
void setup()
{
    Spark.function("digitalread", tinkerDigitalRead);
    Spark.function("digitalwrite", tinkerDigitalWrite);
    Spark.function("analogread", tinkerAnalogRead);
    Spark.function("analogwrite", tinkerAnalogWrite);

    //PUT YOUR SETUP CODE HERE


}

void loop()
{
    //PUT YOUR LOOP CODE HERE


}

int tinkerDigitalRead(String pin) {
    int pinNumber = pin.charAt(1) - '0';
    if (pinNumber< 0 || pinNumber >7) return -1;
    if(pin.startsWith("D")) {
        pinMode(pinNumber, INPUT_PULLDOWN);
        return digitalRead(pinNumber);}
    else if (pin.startsWith("A")){
        pinMode(pinNumber+10, INPUT_PULLDOWN);
        return digitalRead(pinNumber+10);}
    return -2;}

int tinkerDigitalWrite(String command){
    bool value = 0;
    int pinNumber = command.charAt(1) - '0';
    if (pinNumber< 0 || pinNumber >7) return -1;
    if(command.substring(3,7) == "HIGH") value = 1;
    else if(command.substring(3,6) == "LOW") value = 0;
    else return -2;
    if(command.startsWith("D")){
        pinMode(pinNumber, OUTPUT);
        digitalWrite(pinNumber, value);
        return 1;}
    else if(command.startsWith("A")){
        pinMode(pinNumber+10, OUTPUT);
        digitalWrite(pinNumber+10, value);
        return 1;}
    else return -3;}
```

```
int tinkerAnalogRead(String pin){
    int pinNumber = pin.charAt(1) - '0';
    if (pinNumber< 0 || pinNumber >7) return -1;
    if(pin.startsWith("D")){
        pinMode(pinNumber, INPUT);
        return analogRead(pinNumber);}
    else if (pin.startsWith("A")){
        pinMode(pinNumber+10, INPUT);
        return analogRead(pinNumber+10);}
    return -2;}

int tinkerAnalogWrite(String command){
    int pinNumber = command.charAt(1) - '0';
    if (pinNumber< 0 || pinNumber >7) return -1;
    String value = command.substring(3);
    if(command.startsWith("D")){
        pinMode(pinNumber, OUTPUT);
        analogWrite(pinNumber, value.toInt());
        return 1;}
    else if(command.startsWith("A")){
        pinMode(pinNumber+10, OUTPUT);
        analogWrite(pinNumber+10, value.toInt());
        return 1;}
    else return -2;}
```

Using the Particle Web IDE

The Particle Web integrated development environment (IDE) is a simple Web interface that you can use to program your Photon. You can also get useful information and settings about your Photon from here, as well as access tokens for use with the Particle API. To access the Particle IDE, simply head to https://www.particle.io/build. If you have not previously registered for an account, go ahead and create one simply by entering your current e-mail address and password (see Figure 2.7). Registering for an account allows you to save your programs and assign devices to your account. If you have logged into the Particle Build webpage before, then click the Let Me Login button underneath the sign-up button.

Once logged in, you will be presented with the Particle Build programming environment. The Particle Build is an easy-to-use IDE, and this means you can design software programs using software development. The best thing about using the Web IDE is that you can use this on almost any computer that has a Web browser without installing any software programs.

Figure 2.7 *Particle Build login page.*

Particle Build starts with a simple menu bar on the left side of the webpage. At the top of this bar are three buttons, which are the main functions that you will use:

- **Flash** Uploads the current code to the Photon board that is connected. This button initiates the over-the-air firmware update and loads the new software.
- **Verify** This compiles your code before it actually gets flashed and uploaded to the Photon. Just like any software engineering project, you should always compile your code before running it to see if there are any errors. Any errors found will be shown in the debug console at the bottom of the IDE webpage.
- **Save** This simply saves any changes that you have made to your code. When you create a new application and click Save, it is saved to your account and will be accessible every time you log in to the Particle Build IDE.

At the bottom of the menu bar there are four more buttons that allow you to navigate through the Particle Build IDE:

- **Code** Shows a list of applications that you have previously created and saved. This lets you edit and flash the applications to your Photon.
- **Library** Shows a list of applications created by other users and allows you to use and adapt them to your own code. This is a useful

place to look before you start your own applications—after all, why reinvent the wheel?

- **Docs** Brings you all the documentation that is available for the Photon. This is useful if you are stuck or run into issues when using your board.
- **Devices** Shows a list of your Photons or Cores so you can choose which one you want to flash or get information from.
- **Settings** Allows you to change your password, log out from the Web IDE, or get an access token for using the API calls.

For advanced users there are keyboard shortcuts for both Windows and Mac OS X, which can be found on the Github webpage (http://github.com/ajaxorg/ace/wiki/Default-Keyboard-Shortcuts).

On the main Particle Build page you will see the Particle Apps section, which displays the current applications you are working with, as well as a list of any other saved applications you have previously created in the Build IDE and also the community-supported example apps.

Particle Applications and Libraries

When you first open the Particle Build IDE, the application that you will have open in the editor will be the "HELLO WORLD" sample application and has only one file, but the IDE supports multiple associated files, just like the Arduino IDE.

From this main window pane in the IDE, lots of options are available at your fingertips that can help you develop your applications:

- **Create** Allows you to create a new application. It will prompt you to give it a unique name and then press ENTER. Your new app is now saved to your account and ready for editing.
- **Delete** Simply click the Remove App button to remove it forever from your Particle library. Make sure this is absolutely what you want to do, however, as you cannot come back to it.
- **Rename** You can easily rename your application to something cooler if you wish by double-clicking the title of your application under Current App. You can also modify the application description in the same way by double-clicking the description and editing it.

- **My Apps** Under this header you can open up your other applications in a new tab in the editor. This allows you to easily switch between applications and can be useful when you want to copy snippets of code from one app to another.
- **Files** This header lists all of the associated files with your current application. You can click the supporting file to open it up in a new tab and edit it.
- **Examples** The Examples header lists the example apps created by the ever-growing community.

Uploading Your First Application

Now that you have familiarized yourself with the Particle Build IDE, the next best thing to do is just write a simple application so you can understand a bit more how the process of flashing a program to your Photon board works. You will dive into more code later on, but just to give you an idea we will use one of the examples provided with the Particle Build IDE.

First of all make sure your Photon board is connected to the Particle cloud. It should be blinking cyan, which indicates that it's connected to the Particle cloud. In the Particle Build IDE navigate to the Examples header and look for a program called Blink An LED. Click the program, which should now be displayed on the main editor page. Alternatively you can quite easily create a new application and copy the following snippet of code into the active tab. Here is the code that we will use for our first example.

```
//D7 LED Flash Example
int LED = D7;

void setup() {
    pinMode(LED, OUTPUT);
}

void loop() {
    digitalWrite(LED, HIGH);
    delay(1000);
    digitalWrite(LED, LOW);
    delay(1000);
}
```

The next step is to select your Photon to flash your code to. Click the Devices icon in the navigation bar on the left side, and you should see a list of connected devices to the Particle cloud. Next to your Photon name, select the star icon next to it to highlight that device as the active one to program. If you only have one device, this will automatically be selected and you can skip to the next step.

Now that we have our code, click the Flash button, which will send our application firmware to the Photon wirelessly. If the flash was successful, the Photon LED will flash magenta and you should see the on-board blue LED blinking every second. Not the most amazing piece of programming wizardry, but nevertheless you now know the basics of using the Particle's Build IDE environment.

Under the application header you should see a button that says Fork This Example. This allows you to create a copy from the example so you can edit and save your very own version. This is useful when using snippets of code from the community. If you are familiar with using Github, then this function works in the same way as forking a repo.

If you edit the code by changing the value of the `delay()` function from 1000 to 250, which changes the ON and OFF timings of the on-board LED, you need to make sure you hit the Verify button in the navigation menu on the left to confirm that there are no errors when compiling the new firmware. You can then reflash the Photon with the new program, and you should see that the LED is blinking much faster than before.

Account Information

In addition to what we have learned, there are a couple of other features you should become familiar with in the Particle Build IDE, such as viewing important information about your device, managing the Photon associated with your account, and unclaiming your Photon if you want someone else to be able to use your Photon as well.

You can view the Photon's ID by clicking the Devices icon on the left navigation bar and then clicking the drop-down arrow next to the device. If you wish to unclaim your device so someone else can use your board, click your Photon and in the drop-down box click Remove Device. Once the device has been disassociated with your account, it is free to be registered with another user's Particle account.

When you start diving into using the cloud API, at some point you will need to know what your device's API key is. The API key is a unique number registered to your Photon and should be kept secret. Under the Settings tab in your account, you can press the Reset Token button to assign a new API key to your account. Don't forget that if you have any code with your API key already inputted, you will need to change this with the new API key.

Using Libraries

When you want to use code or snippets of code across multiple applications, Particle libraries make it easy for you to accomplish this. Particle libraries are easy to share, with packages built by the community to assist in overcoming those common problems you may encounter when creating your applications. The libraries are hosted on a Web-based service called Github and then easily pulled into the Particle cloud IDE, where they can be included in applications and shared with other users. You can include a library in your application by finding a library that you want to use and clicking the Include In App button, which will in turn add an #include statement to your application so you can use the capabilities of that library.

Adding a library to the IDE requires an open-source Github repository where your code will live on the servers. Github is a Web-based hosting service that offers distributed revision control as well as source code management for users. Github is at the very heart of the open-source community and thrives on users creating software to share. At a minimum this repository requires a spark.json file, some documentation, some example firmware files, and some Arduino/C++ files. The import and validation process is designed to be easy to interpret—you can see an example in Figure 2.8. Just follow the IDE step by step, and it will guide you through what is required to get your library set up and accessible.

The easiest way to generate boilerplate code is to follow these simple steps:

Step 1: Define a function to create library boilerplate.

Copy and paste the following code into a bash or zsh shell or .profile file:

```
create_spark_library() {
    LIB_NAME=$1

    # Make sure a library name was passed
    if [ -z "${LIB_NAME}" ]; then
        echo "Please provide a library name"
```

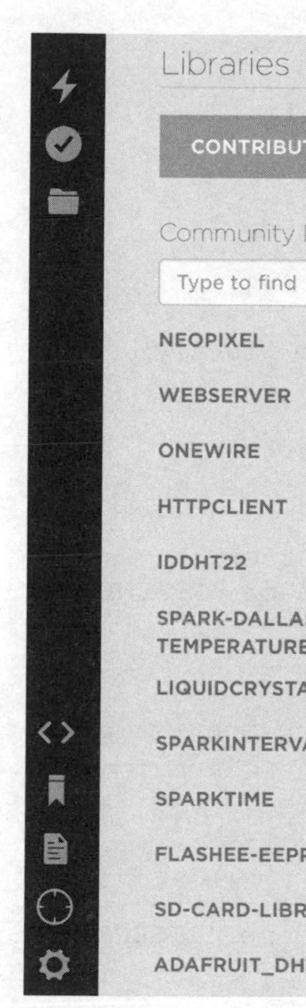

Figure 2.8 *Example library.*

```
        return
    fi

    echo "Creating $LIB_NAME"

    # Create the directory if it doesn't exist
    if [ ! -d "$LIB_NAME" ]; then
        echo " ==> Creating ${LIB_NAME} directory"
        mkdir $LIB_NAME
    fi

    # CD to the directory
    cd $LIB_NAME
```

```
    # Create the spark.json if it doesn't exist.
    if [ ! -f "spark.json" ]; then
        echo " ==> Creating spark.json file"
        cat <<EOS > spark.json
{
    "name": "${LIB_NAME}",
    "version": "0.0.1",
    "author": "Someone <email@somesite.com>",
    "license": "Choose a license",
    "description": "Briefly describe this library"
}
EOS
    fi


    # Create the README file if it doesn't exist
    if test -z "$(find ./ -maxdepth 1 -iname 'README*'
-print -quit)"; then
        echo " ==> Creating README.md"
        cat <<EOS > README.md
TODO: Describe your library and how to run the examples
EOS
    fi


    # Create an empty license file if none exists
    if test -z "$(find ./ -maxdepth 1 -iname 'LICENSE*'
-print -quit)"; then
        echo " ==> Creating LICENSE"
        touch LICENSE
    fi


    # Create the firmware/examples directory if it
doesn't exist
    if [ ! -d "firmware/examples" ]; then
        echo " ==> Creating firmware and firmware/
examples directories"
        mkdir -p firmware/examples
    fi


    # Create the firmware .h file if it doesn't exist
    if [ ! -f "firmware/${LIB_NAME}.h" ]; then
        echo " ==> Creating firmware/${LIB_NAME}.h"
        touch firmware/${LIB_NAME}.h
    fi
```

```
    # Create the firmware .cpp file if it doesn't exist
    if [ ! -f "firmware/${LIB_NAME}.cpp" ]; then
        echo " ==> Creating firmware/${LIB_NAME}.cpp"
        cat <<EOS > firmware/${LIB_NAME}.cpp
#include "${LIB_NAME}.h"

EOS
    fi


    # Create an empty example file if none exists
    if test -z "$(find ./firmware/examples -maxdepth 1
-iname '*' -print -quit)"; then
        echo " ==> Creating firmware/examples/example.
cpp"
        cat <<EOS > firmware/examples/example.cpp
#include "${LIB_NAME}/${LIB_NAME}.h"

// TODO write code that illustrates the best parts of
what your library can do

void setup {

}


void loop {

}
EOS
    fi


    # Initialize the git repo if it's not already one
    if [ ! -d ".git" ]; then
        GIT=`git init`
        echo " ==> ${GIT}"
    fi

    echo "Creation of ${LIB_NAME} complete!"
    echo "Check out https://github.com/spark/uber-
library-example for more details"
}
```

Step 2: Call the function.

```
Create_spark_library this-is-my-library-name
```

Replace `this-is-my-library-name` with your actual library name. It should all be in lowercase letters and dashes between the names.

Step 3: Edit the spark.json firmware .h and .cpp files.
Use the repository as your guide to good library conversions.

Step 4: Create a Github repo and push to it.

Step 5: Validate and publish via the Particle IDE.
To validate, import, and publish the library, go to the Particle Build IDE and click the Add Library button.

Don't worry if you don't understand how to create a library yet, as this is for advanced users. There is plenty of support available on the Particle community site if you're struggling.

Photon's Firmware

The Photon board is a truly embedded device and as such does not have an operating system like other traditional computers have. Instead, it runs on a single piece of code called firmware, which runs whenever the device is powered on.

Traditional hardware has hard-coded software embedded in and thus it is difficult to change or flash new firmware every time. The Photon board uses over-the-air firmware updates to overwrite almost all of its software with a new one. The only piece of software that is not affected when flashing the device is the bootloader, which manages the process of uploading new firmware and ensures a successful load. The bootloader is also responsible for the factory reset option that is provided on the Photon.

Summary

That concludes this chapter for now—hopefully you are able to connect your Photon to the Wi-Fi network and are comfortable creating your very own applications using the Particle Build IDE webpage. In the next chapter we will look at some of the Arduino-style C programming concepts to understand a bit about how we can program the Photon board.

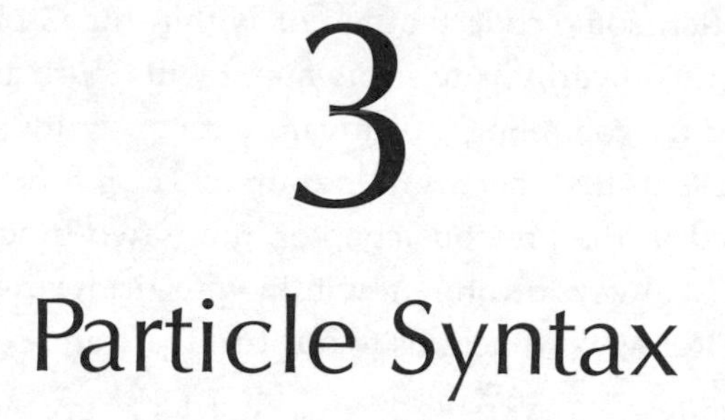

3 Particle Syntax

The programming language used to program the Photon is called C. In this chapter you will learn and understand some of the basic programming terms using this language. You can use what you will learn here and apply this to most of the firmware you will write throughout the book. To get the best out of using your Photon, you will need to learn these basic programming fundamentals.

What Is Programming?

It may not seem obvious to a beginner what programming actually is and what a programming language is. When you look at the Photon's firmware you could probably hazard a guess as to what it is actually doing without any programming knowledge, but we need to look a bit further into how the code goes from being lines of text to something in real time, like turning a light-emitting diode (LED) on or off.

When you press the FLASH button on the Particle Build integrated development environment (IDE), it then implements a chain of events that results in your firmware being uploaded to the Photon and run. What it actually does is something called compilation, where it takes your lines of code as text and translates them into something called binary, which is a series of 1's and 0's that that the Photon's hardware will understand. Recall from the previous chapter that you clicked the verify button before you actually flashed any of your code to the Photon. This attempts to precompile the C code that you have written without actually flashing it. Verifying your code also makes sure that what you have written makes sense in the C programming language.

If you have written some code that is not within the C programming language, then when you verify your firmware, it will return an error. The same is true when you try compiling a firmware with no written code at all—the error returned tells us that there is no `setup` or `loop` function in your code. As we mentioned in the previous chapter, these two functions of code are required and must always be present within your firmware.

Let's add the following functions to our firmware and see if it compiles:

```
void setup (){
}
void loop() {
}
```

When verifying the firmware the compiler will tell you that it has successfully compiled your code and that everything was acceptable to the C language standard. At the bottom of the compile if you click the little icon for information, it will also tell you how much of the flash memory has been used.

The Photon has a total of 1MB internal flash memory, which is divided into three main areas. Beginning at the top of the memory space is where the bootloader is saved and locked. The second region is reserved for storing system flags, and the third region holds the actual user firmware.

Let's take a closer look at our `setup` and `loop` functions that will always be the starting point of every firmware that we write. We can start by using the word `void` before `setup` and `loop` followed by a pair of curly braces.

The line `void setup ()` means that we are defining a function called setup within our code. Some functions are already defined for us, such as `digitalWrite` and `delay`. `Setup` and `loop` are two functions that we must define for ourselves in every program we write.

We are not calling `setup` or `loop` like we do with `digitalWrite` or `delay`, but we are actually creating these functions so that the Photon itself can call them. This might sound a little bit confusing, but the best way to think of it is that we are trying to shorten our code. Using `void` with both `setup` and `loop` allows us to not return any values within the function, unlike other functions, so we have to say that these are void.

After the word `void` comes the function's name and parentheses containing any arguments. In our case both `setup` and `loop` do not contain any arguments, but we still have to include the parentheses. Because we are defining a function within our code and not calling a function, we do not have to close it with a semicolon. Instead, we use the curly braces, which is where our code sits between in the function—this is known as a block of code. Just because we define the functions `setup` and `loop` that does not necessarily mean we have to use those functions to hold any block of code—we simply just need to define them in every firmware we write, although in reality this may not ever happen.

Let's go back to our example sketch in Chapter 2:

```
//D7 LED Flash Example
int LED = D7;

void setup() {
    pinMode(LED, OUTPUT);
}

void loop() {
    digitalWrite(LED, HIGH);
    delay(1000);
    digitalWrite(LED, LOW);
    delay(1000);
}
```

The `setup` function in our sketch calls one built-in function called `pinMode`. The function `pinMode` is used to set a particular pin to be either an input or an output. So with this in mind it is clear that we need to initially set our LED to be an output, which will also let us use the function `digitalWrite` later on. The `pinMode` is always used in the `setup` function because we only need to set the pin mode once in our sketch. The program would still work if you called this in the `loop` function, but for best coding practices it is always best to keep things that you call once in the `setup` function—then you know where everything is that you only call once.

Variables

A variable is a place in the memory to store a piece of data. It has a name, a value, and a type. For example, the following statement declares the pin number:

```
int pin = D0;
```

This code creates a variable called pin whose value is D0, and its type is `int`. Later on in your program you can refer to this variable by its name, at which point its value will be looked up and used. For example:

```
pinMode(pin, OUTPUT);
```

It is the value of the pin that will be passed into the `pinMode()` function. In this particular case you don't actually need to use a variable; this statement would work just as well if you referenced the pin number directly:

```
pinMode(D0, OUTPUT);
```

The advantage of a variable in this case is that you only need to specify the actual number of the pin once, but it has a huge advantage in that you can use this more than once in your code. When naming your variable you can also use a descriptive name, which will make the significance of the variable clearer (e.g., a program controlling some LEDs might have variables called redPin, GreenPin, etc.).

A variable has other advantages over a value like a number. You can change the value of a variable using an assignment (indicated by an equal sign). For example:

```
pin = D1;
```

This will change the value of the variable to D1. You can see that we don't actually specify the type of variable when changing the value—only the name of the variable and the variable type stay constant. Remember that you have to declare a variable before you can assign a value to it. You can, of course, make a copy of a variable within your code if needed. When you change the value of one variable, it does not affect the value in another. This is useful for when you change the value of a variable but also want to keep the original value in case you need to go back to it. For example:

```
int pin = D0;
int pin2 = pin;
pin = D1
```

Only the variable pin would have changed to D1, and pin2 would remain the same (D0).

If you try changing the value of a variable before you have declared the variable, you will receive an error in your code when trying to compile it: "error: pin was not declared in the scope." (Scope refers to the part of your program where the variable can be used.) This is determined by where you declare it. For example, if you want to be able to use your variable anywhere within your program, then you must declare the variable at the top of the program. This is known as a *global* variable. Here is an example of declaring a global variable:

```
int pin D0;
void setup() {
      pinMode(pin, OUTPUT);
{
void loop() {
      digitalWrite(pin, HIGH);
}
```

As you can see in the example, pin is used in both the `setup` and `loop` functions. Both functions are referring to the same variable; therefore, it must be set as a global variable. If you only need to use a variable in a single function, then you declare it there, in which case its scope will be limited to that particular function. For example:

```
void setup() {
      int pin D0;
      pinMode(pin, OUTPUT);
      digitalWrite(pin, HIGH);
}
```

In this example the variable pin can only be used inside the `setup` function. If you try to use it within the `loop` function, you will receive an error in your program. You may be wondering why we don't simply declare all variables as global variables at the start of the program. Well, it makes it easier to find out what has happened to the value of the variable. If you remember that when we use a global variable the value can be changed anywhere within the whole program, this means you need to understand your program in order to know what happened to the variable. Sometimes it is much easier to debug when you only use that variable within its own scope.

Floats

All the examples that we have used thus far have included `int` variables. Integer variables are by far the most commonly used type; however, there are others that you should be familiar with.

One type that will become more relevant later on in this book is `floats`. A good example is the conversion of temperature when using temperature sensors. The variable type is a number that may consist of a decimal place for more precise measurements such as 1.6. Take a look at the following formula:

f = c * 9 / 5 + 32

This formula is the calculation for converting temperature from degrees Celsius to degrees Fahrenheit. If we give c the value of 23, then f will be 23 * 9 / 5 + 32 or 73.4. If we set f as an integer, then the value will return 73.

Note the order in which the calculation is in. If we are not careful in the order in which we calculate, things may turn out differently when using integers. For example, this formula:

f = (c / 5) * 9 + 32

would result in the following:

- 23 is divided by 5, which returns 4.6, which is then rounded down to 4.
- 4 is then multiplied by 9 and 32 is added to give a result of 68, which is way off our actual temperature value of 73.4.

For situations like this, we use floats; in the following example our temperature conversion function is rewritten to use floats:

```
float centToFaren (float c)
{
        float f = c * 9.0 / 5.0 + 32.0;
        return f;
}
```

We have also added .0 to the end of each value—this way our compiler knows that it should treat the values as floats and not integers.

Boolean

Another common type of variable is a Boolean, which has logical values—that is, the value is either `true` or `false`. The best example of using Boolean logic is using the if statement. The condition set in an if statement can only be `true` or `false`.

```
int LEDpin = 5;         // LED on pin 5
int switchPin = 13;     // momentary switch on 13, other side connected to ground

boolean running = false;

void setup()
{
  pinMode(LEDpin, OUTPUT);
  pinMode(switchPin, INPUT);
  digitalWrite(switchPin, HIGH);      // turn on pullup resistor
}
```

```
void loop()
{
  if (digitalRead(switchPin) == LOW)
  { // switch is pressed - pullup keeps pin high normally
    delay(100);                         // delay to debounce switch
    running = !running;                 // toggle running variable
    digitalWrite(LEDpin, running)       // indicate via LED
  }
}
```

In addition, you can manipulate values using Boolean operators. These are similar to performing arithmetic calculations. The most commonly used Boolean operators are `and`, which is written as `&&`; and `or`, which is written as `||`. In addition to these operators there is `not`, which is written using !. This value can either be "not true" or "not false." Table 3.1 shows all the data types that are available and which format of values should be used.

Another thing to remember is what happens when your values exceed the limits. This causes odd things to happen. For example, if you have the byte variable with a value of 255 and you suddenly add a 1 to this value, it returns a zero; similar to this, if you add a 1 to an integer with the maximum value of 32767, it becomes a negative value of –32768. Usually, you can get away with most of your data types being an integer, so sometimes it's best to use this data type by default.

Char

The data type `char` is a byte that represents an *ASCII* character. ASCII (American Standard Code for Information Interchange) is a system from the very early days of computing used to translate between bytes and characters.

Data Type	Memory (bytes)	Range
boolean	1	true or false (0 or 1)
char	1	–128 to +128
byte	1	0 to 255
int	2	–32768 to +32767
unsigned int	2	0 to 65536
long	4	2,147,483,648 to 2,147,483,647
unsigned long	4	0 to 4,294,967,295
float	4	–34028235E+38 to +3.4028235E+38
double	4	same as float

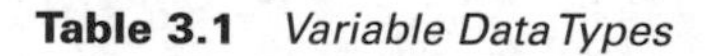

Table 3.1 *Variable Data Types*

A char typically only takes up 1 byte of memory that stores a character value. Single characters are written in single quotes 'A' and multiple characters are written in double quotations "ABC." In theory, a char is stored as a number according to the ASCII table (e.g., A is equal to the number 65). Here is how you create and assign a char variable:

```
char letter = 'A';
char letter = 65;
```

Both examples of using char are correct.

Commands

The C language on the Photon has a number of built-in commands. In this section we will explore some of these commands and see how they can be used in our firmware.

The if Statement

In our examples so far we have assumed that your lines of programming will be executed in order one after the other. But what if we don't want to do this and we want to execute a block of code when something happens in the code. For this we can use an if statement, which is used in conjunction with a comparison operator, and test whether a certain condition has been reached, such as an input being higher or lower than a certain number. The formatting for an if statement is as follows:

```
if (variable > 50)
{
//Write your code here
}
```

The program tests to see if the variable is greater than 50. If it is, then the program takes a particular action and executes the code in between the curly braces. If the condition is not reached, then the program skips over this section and moves on to the next. Also, the curly braces may be omitted after an if statement; if this is done, then the next line of code becomes the only conditional statement, as shown in the next example:

```
if (x > 50) digitalWrite(LEDpin, HIGH);
     if (x > 50)
     digitalWrite(LEDpin, HIGH);
```

```
if (x > 50) { digitalWrite(LEDpin, HIGH); }
if (x > 50) {
      digitalWrite(ledPin1, HIGH);
      digitalWrite(ledPin2, HIGH);
}
```

All of the examples are correct for using an `if` statement. Note that we used the symbol >, which means "more than." It is one example of what are called comparison operators. These operators can be found in Table 3.2.

Just remember that when using the equal sign you must use a double equal symbol, which is a comparison operator, rather than a single equal symbol, which is an assignment operator. It is easy to get the two mixed up when using conditions. If we accidently used a single equal operator, then the `if` statement would always return a true condition. This is because the C language always evaluates the statement as an assignment, so in our example, x would be set to 50 and would always be true.

There is another form to the `if` statement that allows you to do another thing if the condition has not been reached or it is false. We will use this in some practical examples later on in the book.

for Loops

You may find yourself wanting to run a series of commands a number of times within your programs. We know already that we can use the `loop` function—when all the lines of code in the `loop` function have run, it will just start over again at the beginning. This is great—but we only want to run our code a few times to get a certain desired result. For this we can use something called a `for` statement, which is used to repeat a block of code

Operator	Meaning	Examples	Result
<	Less than	9 < 10 9 < 9	True False
>	Greater than	10 > 9 10 > 10	True False
<=	Less than or equal to	10 <= 10 9 <= 10	True True
>=	Greater than or equal to	10 >= 10 10 >= 9	True True
= =	Equal to	9 == 9	True
!=	Not equal to	9 !=9	False

Table 3.2 *Comparison Operators*

within the curly braces. Usually you would use an increment counter to determine how many times you want to loop the code. The `for` loop is useful for any kind of repetitive operation and is often used in combination with arrays. There are three main parts to the `for` loop header:

```
for (initialization; condition; increment) {
         statements
}
    for (int x = 0; x < 10; x++) {
         delay(500);
    }
```

The `initialization` happens first and only once; each time through the loop, the `condition` is tested. If it is true, then the statements are executed, as well as the `increment`, and then the condition is tested again. The condition becomes false when the loop has repeated more than 10 times.

while Loops

Another useful way of looping in C is to use the `while` command in place of the `for` command. `while` loops will loop continuously until the expression inside the parentheses () becomes false. Something has to change the tested variable; otherwise, the `while` loop will never end and will run indefinitely. The syntax for using a `while` loop is as follows:

```
while(expression) {
     statements
}
int i = 0;
while (i = < 10)
{
     delay(500);
     i ++;
}
```

The expression in the parentheses after the `while` must be true to stay in the loop. When it is no longer true, then the sketch continues running the commands after the curly braces. You may also notice the following line:

```
i ++;
```

This is just C shorthand for the following expression:

```
i = i + 1;
```

Arrays

Arrays are a way of containing a list of different values. This is unlike what we have learned before, where variables contained only a single value, usually an `int` data type. In contrast to this, an array contains a list of values, and you can easily access any one of those values by its particular position in the array. In most programming languages, and in fact computer science in general, the first value is always represented as a 0 rather than a 1; this means that the first variable is actually element zero. Here are some examples of how to declare an array in your code:

```
int myValue[6];
      int myPins[] = {2, 4, 8, 3, 6};
      int mySensVals[6] = {2, 4, -8, 3, 2};
      char message[6] = "hello";
```

You can declare an array without initializing it, as in `myValue`. In `myPins` we declare an array without explicitly choosing a size. The compiler counts the elements and creates an array of the appropriate size. Finally you can both initialize and size your array, as in `mySensVals`. Note that when declaring an array of type `char`, one more element than your initialization is required to hold the required null character.

Arrays are zero indexed—that is, referring to the array initialization earlier, the first element of the array is at index 0, hence `mySensVals[0] == 2`, `mySensVals[1] == 4`, and so forth.

It also means that in an array with 10 elements, index nine is the last element. Hence:

```
int myArray[10]={9,3,2,4,3,2,7,8,9,11};
     // myArray[9]    contains 11
     // myArray[10]   is invalid and contains random
information (other memory address)
```

For this reason, you should be careful in accessing arrays. Accessing past the end of an array (using an index number greater than your declared array size) is reading from memory that is in use for other purposes. Reading from these locations is probably not going to do much except yield invalid data. Writing to random memory locations is definitely a bad idea and can often lead to unhappy results, such as crashes or your program malfunctioning. This can also be a difficult bug to track down.

Strings

A string is a sequence of characters and a way for your Photon to deal with text. It is highly unlikely that we would use strings within our code—possibly if you are using a liquid crystal display (LCD) display, then a string might come into play.

All of the following are valid declarations for strings:

```
char Str1[15];
char Str2[8] = {'a', 'r', 'd', 'u', 'i', 'n', 'o'};
char Str3[8] = {'a', 'r', 'd', 'u', 'i', 'n', 'o', '\0'};
char Str4[ ] = "arduino";
char Str5[8] = "arduino";
char Str6[15] = "arduino";
```

Possibilities for declaring strings include the following:

- Declare an array of `chars` without initializing it, as in `Str1`.
- Declare an array of `chars` (with one extra `char`), and the compiler will add the required null character, as in `Str2`.
- Explicitly add the null character, `Str3`.
- Initialize with a string constant in quotation marks; the compiler will size the array to fit the string constant and a terminating null character, `Str4`.
- Initialize the array with an explicit size and string constant, `Str5`.
- Initialize the array, leaving extra space for a larger string, `Str6`.

Generally, strings are terminated with a null character: ASCII code 0. This allows functions to tell where the end of a string is. Otherwise, they would continue reading subsequent bytes of memory that aren't actually part of the string.

This means that your string needs to have space for one more character than the text you want it to contain. That is why `Str2` and `Str5` need to be eight characters, even though it is only seven—the last position is automatically filled with a null character. `Str4` will be automatically sized to eight characters, one for the extra null. In `Str3`, we've explicitly included the null character (written '\0').

It is possible to have a string without a final null character (e.g., if you had specified the length of `Str2` as seven instead of eight). This will break most functions that use strings, so you shouldn't do it intentionally. If you notice something behaving strangely (operating on characters not in the string), however, this could be the problem.

Coding Best Practices

The Particle compiler does not pay any attention to how you lay out your code, but it does require you to write all your code on a single line with a semicolon between each statement. If you think about how you may read a book, usually the formatting is very similar—you have your table of contents, chapters, paragraphs, and index.

Formatting code is a personal choice—some like to keep it messy and some like to keep things neat and tidy with additional commenting between sections. Usually keeping code tidy is always best—it makes decoding bugs quicker and allows someone else to easily read your code if you collaborate with others.

Indentation

You can see in some of the example sketches that we always use some sort of indentation for the code in the left margin. Indentation is usually determined by the curly braces and provides some sort of hierarchy structure to the whole code. Using the following example we can see that we have `void loop()` as our top level and within this we have a small amount of code followed by another sublevel starting with `if`:

```
Void loop()
{
        int count = 0;
        count = ++;
        if  (count == 10)
                {
                        count = 0;
                        delay(1000);
                }
}
```

If we added another `if` statement within the first, then we would simply increase the indentation by a further 1 or 2. To indent from the left margin, you can simply press TAB on the keyboard to move across. You might find this a bit tedious, but when reviewing your code later on it will become apparent how useful this is.

Commenting Your Code

Comments in your code is text that the compiler does not read and simply ignores. Comments can provide additional information to either you, as the programmer, or someone who is reading your code. If your program has a lot of code divided into many sections, then you can also use comments as titles or headers; this can be useful when debugging code so you can easily find the section you need to edit or change. There are currently two forms of syntax in which you can write comments in your code:

- The single-line comment, and probably the most common one that you will use, starts with two backslashes together // until the end of the line. So you cannot insert a comment and then carry on with code on the same line, as the code will be ignored by the compiler because it thinks it is part of the comment.
- The multiline comment is separated by /* and */. You can use this at the start of your program to introduce the code and write a short description of what your program does.

The following example shows both types of commenting syntax:

```
/* This is an example of how to write different types of
comments with your program.
Written by Christopher Rush
*/
void loop() {
      int count = 0;
      count ++; //adds plus one to the integer count
      if (count == 10) {
            count = 0;
            delay(1000) //pauses for 1 second
      }
}
```

In this book I will stick to using single-line comments, usually to help explain what is happening in the code. It is useful if other people are going to use the code or snippets of the code in their projects. Sometimes it can be confusing for the beginner knowing when and when not to use comments; however, I usually follow a few simple rules that should make things a bit easier. Comments should be used to

- Explain anything that can be a little tricky to comprehend

- Describe something the user may need to do that is not written in the code; for example, //LED must be wired up to pin D1
- Leave yourself notes or instructions: // Note: tidy up this code with an easy function

The last point can be very useful to use either Note or Todo, which reminds you that you need to come back to this point sometime in the future. Some IDE compilers allow you to search for keywords.

Whitespaces

The compiler program will always ignore any whitespace lines in your program, unless they are spaces that separate words in your code—that is, the following code will still work, but reading it or debugging would prove to be very difficult:

```
void loop() {int
count = 0; count ++; if(
count==10) {count =0;
delay(1000);}}
```

Some users tend to put space between everything, and others try to create a more streamlined format—either way it doesn't matter; the compiler will still read the code in the way it is intended to.

```
int count = 0;
int count=0;
```

Summary

I know that some of you just want to get down to the nitty gritty of the Photon, but it is important to understand the basic concepts of programming the Photon board. This has largely been a theoretical chapter. You have learned how to write your firmware so that it is easy to read and understand, as well as adopted the Arduino style of programming language, which will save you a lot of time and effort when writing your own firmware for the Photon.

In this next chapter we will look at programming some electronics using output devices such as LEDs, motors, and servos for both analog and digital circuits.

4

Outputs

This chapter introduces you to controlling output devices such as light-emitting diodes (LEDs), relays, and buzzers. Output devices are usually used to communicate information, such as the status of a circuit, or to switch something on or off, such as a direct current (DC) motor or servo. The Photon and Core are all about connecting physical devices to the world, and this means connecting electronic components to your Particle board. Outputs on the Photon are digital, which means switching between 0 V and 3.3 V. Outputs can also be analog signals, which allow you to set a varying voltage to a device between 0 and 3.3 V, although in reality it is not as simple as it may seem. This book is primarily about software programming rather than the hardware side of things, however, so let's not get too dragged down into the complexity of the circuit, but rather focus on the programming. Understanding the basic principles of the circuit will help you know what is happening and why.

Digital Outputs

The photon board has a whole host of pins available from D0–D7 and A0–A5. All these pins, by default, are output pins, but we can configure them in such a way in our firmware that they become output pins and can control output devices.

To understand how the digital output pins work, there is a simple experiment that we can try on one of the digital pins on the Photon board. This experiment involves the use of a basic digital multimeter and some prototyping wire, as set out in Table 4.1.

Description	Appendix
Photon board	M1
Breadboard	H1
Jumper wires	H2
Digital multimeter	H3

Table 4.1 *Components and Hardware*

Let's use digital pin 0 for this experiment to see what's going on. Insert the Photon board into the breadboard, making sure that the pins are inserted on either side of the bridge in the middle to prevent any shorting out, which could potentially damage the board. Insert one jumper wire into the breadboard hole next to digital pin D0 and insert the another jumper wire next to the GND pin, which will be used to complete the circuit. Figure 4.1 shows the arrangement of the pins and jumper wires. If your multimeter has crocodile clips, make sure the jumper wires have bare ends to create a contact with the crocodile clips and attach the clips to the end. Chances are that your multimeter does not come supplied with crocodile clips, which isn't a problem—you can simply strip the jumper wires back a bit more and wrap the wire around the end of the probe. If necessary, affix some electrical tape to keep everything secure.

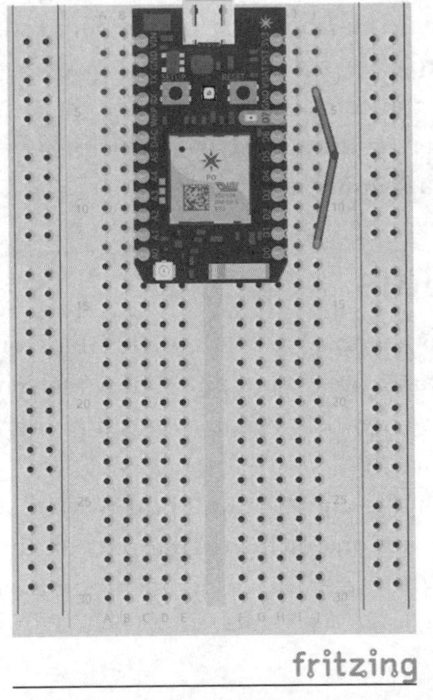

Figure 4.1 *Measuring digital outputs on the Photon.*

Set the multimeter range to somewhere between 0 and 20 V DC, as we already know that the Photon has an output of 3.3 V. The negative lead (black) on the multimeter should always be connected to the GND pin on the Photon board and the positive lead (red) to the digital pin D0 on the Photon board.

Here is the firmware that we will load to the Photon board:

```
int digitalpin = D0;

void setup() {

  pinMode(digitalpin, OUTPUT);

}

void loop() {

  digitalWrite(digitalpin, HIGH);

  delay(1000);

  digitalWrite(digitalpin, LOW);

  delay(1000);

}
```

At the start of the program in the `setup` and `loop` functions you can see the command `pinMode`. You use this command every time you are using a pin on the Photon, whether this is an input pin or an output pin. This allows the Photon board to configure the electronics connected to that pin and allows us to control them using some simple commands in the firmware.

As you may already know from the previous chapters, certain functions are built-in, and pinMode is one of those functions. Its first argument is the pin number that the function is referring to. This pin number is an integer value represented by a number. The second argument is the mode, which must either be an INPUT or OUTPUT.

NOTE *The mode value must always be uppercase.*

The loop part of the program switches the digital pin on the Photon board to HIGH and waits for one second in the delay and then sets the pin back to LOW and waits another second before repeating the process.

So with the multimeter turned on and plugged into the Photon board, you should be able to see its readings change from 0 V to 3.3 V as the firmware

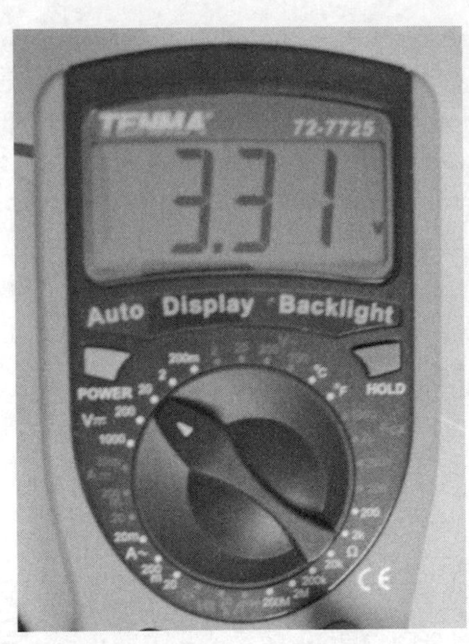

Figure 4.2 *Setting the output to HIGH.*

runs, as shown in Figure 4.2. This is a simple way of showing how the digital pins work as they send the voltage from 0 V to 3.3 V using the command DigitalWrite and setting the argument to HIGH or LOW.

With the following in mind, you can easily see how we can use the Photon board to control electronic devices and components using some simple commands and basic circuitry.

Flashing an LED

Moving on from the previous experiment, we will use the same principle but this time will be creating a circuit with an LED and a resistor so we can flash the LED on and off.

LEDs will definitely be one of the most commonly used parts in your projects throughout this book. They are inexpensive and easy to obtain from your local electronics store (see Appendix A for a list of suppliers). LEDs are polarized, which means that it matters which way you connect it to your circuit. The positive leg on the LED is called the anode, and the negative leg on the LED is called the cathode. If you look at your LED, on the top of the plastic shell you can usually see a flat side to the casing. This side is the cathode. Another simple way to determine which side is anode and which

Schematic Reference	Description	Appendix
M1	Photon board	M1
	Breadboard	H1
	Jumper wires	H2
D1	5-mm LED	S1
R1	220-ohm resistor	R1

Table 4.2 *Components and Hardware*

is cathode is by taking a look at the length of the legs of the LED. The longest leg is always the anode and the shortest is the cathode.

With all types of LEDs, the current only flows in one direction: from the anode to the cathode. As a result, the anode should always be connected to power source. In our instance, this will be the voltage output from the digital pin on the Photon board, which is 3.3 V. When using LEDs, it is also common to see them run in series with a resistor. Resistors are not polarized; therefore, you do not need to worry about how they are connected to the circuit. Table 4.2 shows the components we will be using for this experiment.

You will need to connect the LED to pin D0 on the Photon board; you can do this by inserting the LED into the breadboard as shown in Figure 4.3.

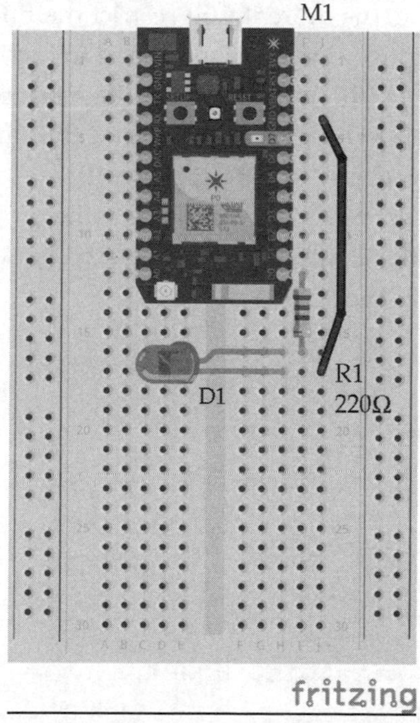

Figure 4.3 *Breadboard layout diagram for LED circuit.*

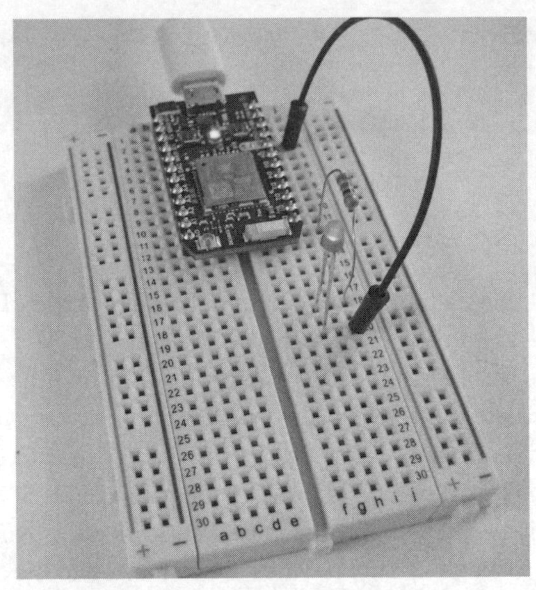

Figure 4.4 *Photon wired to an LED.*

Don't forget that we also need to connect a resistor in series to act as a current limiter, which will prevent damage to the LED or the Photon board. The larger the resistor value, the more it will restrict the flow of current and the dimmer the LED will glow. Connect the resistor to pin D0 and the other end to the anode of the LED (the longest leg). Take a jumper wire and connect the cathode leg of the LED to the GND pin on the Photon board (refer to Figures 4.3 and 4.4 to see how to wire the circuit).

For this experiment we are going to use the same program we used earlier when we tested the voltage outputs using the `DigitalWrite` function to set the output to either HIGH or LOW.

```
int digitalpin = D0;

void setup() {

  pinMode(digitalpin, OUTPUT);

}

void loop() {

  digitalWrite(digitalpin, HIGH);
```

```
    delay(1000);

    digitalWrite(digitalpin, LOW);

    delay(1000);

}
```

Load this program into your Photon and see what happens. The LED will flash on for one second and wait another second before switching off. This is a simple program that shows the basics for outputs using the Photon board.

LCD Display

One of the best things about using an embedded device such as the Photon board is that it can operate independently from any computer. But if you want to display information, it can be rather tricky using embedded boards, and sometimes a simple traffic light LED system just does not display information. A commonly found component on finished goods is a liquid crystal display (LCD). This can be used to display more complex information such as sensor values, timing information, settings or progress bars, etc. In this experiment you will learn how to connect a standard LCD display to the Photon board and how to program it to display any information that is relevant to your projects. We will also look at already existing libraries that can be found using the build environment, and we can import those into our projects, which can save some time when writing our program.

To complete the example in this chapter we are going to use a parallel LCD screen. These are common LCD displays and often can be salvaged from old electronic finished goods that have been lying around collecting dust in your basement for years such as old VCRs or DVD players. They come in all different kinds of shapes and sizes, with the most common being a 16 × 2 character display with a single row of 14 pin headers or 16 if it has a backlight. This configuration gives us a maximum of 32 characters we can display at any given time, with 16 characters on two rows.

If your LCD display did not come with any pin headers already soldered on, you will need to solder one on so you can insert it into the breadboard and then use jumper wires to connect the Photon board. With the headers successfully soldered on to the LCD display, it should something like Figure 4.5.

Now that we have our LCD display ready, the next part involves connecting it to the Photon board, which can be a little bit difficult and maybe even

Figure 4.5 *LCD display with headers soldered on.*

confusing. All the standard LCD displays have the same pin-out functions and can be wired in two different modes: four-pin and eight-pin mode. For this example we can simply wire up the LCD display in four-pin mode, which will give us enough options to complete our project and allow communication between the display and the Photon board. There are also pins for enabling the display, setting the display to command mode or character mode, and for setting the read/write function. Table 4.3 shows each of the pins and their functions.

Pin Number	Pin Name	Pin Function
1	VSS	Ground
2	VDD	+5V
3	V0	Contrast adjustment
4	RS	Register selection
5	RW	Read/write
6	EN	Enable
7	D0	Data line 0
8	D1	Data line 1
9	D2	Data line 2
10	D3	Data line 3
11	D4	Data line 4
12	D5	Data line 5
13	D6	Data line 6
14	D7	Data line 7
15	A	Backlight anode
16	K	Backlight cathode

Table 4.3 *Parallel LCD Pins*

Here is a bit more detail on the breakdown of the pin connections; most of this information can be found on the manufacturer's datasheet for that particular type of LCD display:

- The contrast adjustment pin changes how dark the display is. This connects to a center pin on a potentiometer so you can easily adjust it on the fly.
- The register selection pin sets the LCD to command or character mode. This is used so the LCD display knows how to interpret the next set of data that is transmitted through the data lines. Based on the state of that pin, the data sent to the LCD display is either a command or a character.
- The RW pin is always connected to the GND pin because we are only ever writing to the LCD display and not reading the data.
- The EN pin is used to tell the LCD display that you are ready to send data.
- Data pins 4–7 are used for actually transmitting data, whereas data pins 0–3 are left unconnected.
- Some LCD displays have backlit illumination in various colors. You can connect this just like you would an LED to the Photon, using the anode and cathode pins with a resistor in series.

You can connect the data pins on the LCD display to any of the digital output pins on the Photon board. Usually, it's best practice to keep them all together and in a particular sequence. For this experiment they are connected to the Photon board as shown in Table 4.4.

Now that you know what pins we need to connect to the Photon board, you can go ahead and wire up the LCD display according Figure 4.6; the hardware required for this experiment can be found in Table 4.5.

LCD Pin	Photon Pin
RS	D0
EN	D1
D4	D2
D5	D3
D6	D4
D7	D5

Table 4.4 *Data Pin Connection to the Photon Board*

Schematic Reference	Description	Appendix
M1	Photon board	M1
	Breadboard	H1
	Jumper wires	H2
H1	16 × 2 LCD display	H4
R1	10-K potentiometer	R2

Table 4.5 *Components and Hardware*

NOTE *If your LCD display requires a 5-V backlight, you can connect this to the VIN pin on the Photon board, which references the input voltage from the universal serial bus (USB).*

Now that the LCD display is ready, we can get the code ready and start displaying some text on the screen. The potentiometer will be used to adjust the contrast between the text and the background so it will be easier to read.

The Particle integrated development environment (IDE) includes the *LiquidCrystal* library, which is a set of predetermined functions that make it easy to interface with any parallel LCD display that you are using. The LiquidCrystal library has an impressive amount of functionality, including

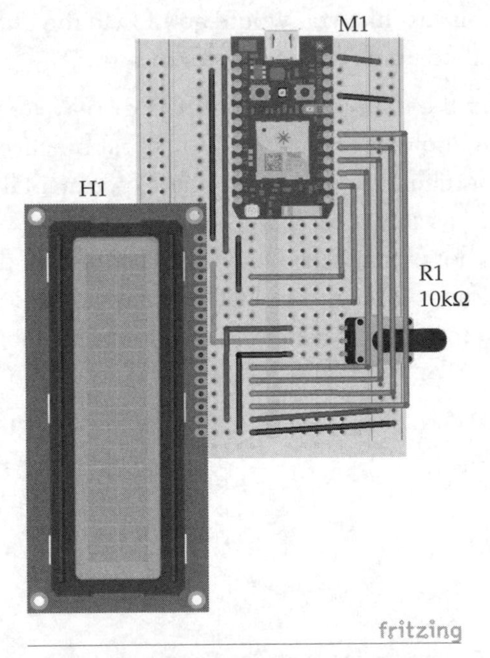

Figure 4.6 *Breadboard layout diagram for LCD display.*

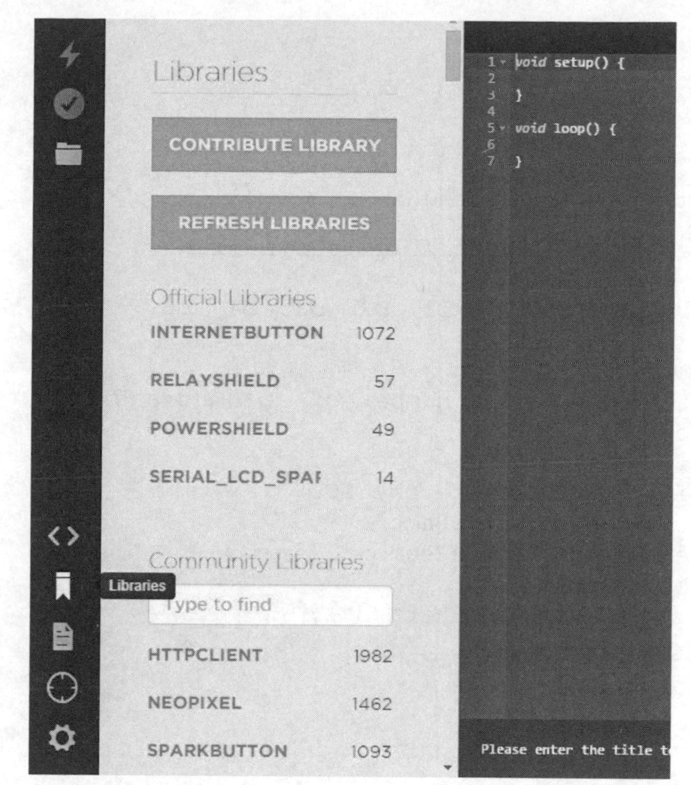

Figure 4.7 *Particle libraries.*

blinking the cursor, scrolling text, creating custom characters, and changing the direction of the printing. This experiment does not show all these features, but rather gives you the tools you need to understand how to interface with the LCD display using the most commonly used functions and features.

To import this library into our program, you need to navigate to the Libraries tab on the left side as shown in Figure 4.7. Type "liquidcrystal" in the search box or, alternatively, find the library from the top list of forked libraries.

Open up the LiquidCrystal library, click the Include In App button, and select which program you want to include the library in. If you haven't done so already, create new blank program and save it as "LCD Display" or whatever you want to call the program. Once this is done, you should see the following lines added to your program:

```
#include "LiquidCrystal/LiquidCrystal.h"

void setup() {
```

```
}

void loop() {

}
```

The rest of program is as follows:

```
#include "LiquidCrystal/LiquidCrystal.h"

LiquidCrystal lcd(D0, D1, D2, D3, D4, D5);

void setup() {
  // set up the LCD's number of columns and rows:
  lcd.begin(16,2);
  lcd.clear();
  // Print a message to the LCD.
  lcd.print("Particle Photon");
  //Move cursor to the second line
  lcd.setCursor(0, 1);
  lcd.print("Getting Started");
}

void loop() {

}
```

In this example you add some text to the display on the first and second lines of the LCD display. This demonstrates how to initialize the display, how to write text, and how to move the cursor. First include the LiquidCrystal library as previously mentioned:

```
#include "LiquidCrystal/LiquidCrystal.h"
```

Then initialize an LCD object as follows:

```
LiquidCrystal lcd(D0, D1, D2, D3, D4, D5);
```

The arguments for the LCD initialization represent the Photon pins connected to RS, EN, D4, D5, D6, and D7 in that order. In the setup you call the library's `begin()` function to set up the LCD display with the character size. The arguments for this command represent the number of columns and the number of rows:

```
lcd.begin(16,2);
```

After this setup procedure you can now call the functions `print()` and `setCursor()` commands to print any given text to any particular place on the LCD display. The positions on the screen are indexed starting with (0,0) in the top left of the screen. The first argument `setCursor()` specifies the column number, and the second specifies the row number. By default, the starting location is always (0,0). So if you call the first command `print()` without first changing the cursor position, it will print starting in the top-left corner.

NOTE *The library does not check to see if the text you are printing fits in the display, so make sure your text is within the 16-character limit.*

This experiment is a great example of how to display information using an LCD display. The display can also show information from sensors, such as temperature and humidity, as well as date and time stamps. You should now be able to experiment with displaying other information.

Analog Outputs

Now that we have looked at digital output controls, it is time to understand a bit about the analog pins on the Photon board. Although controlling digital components is good, what if we wanted to control them with a bit more precision, such as the brightness of an LED or the speed of a motor? Well, you can't normally do this using a digital system, but there is a way using a simple digital-to-analog convertor (DAC) chip and using something called pulse width modulation (PWM).

Pulse Width Modulation

First let's take a look at using PWM, which is commonly found on most single-board computing devices such as the Photon. This method allows you to get pretty close to generating analog values by emulating what an analog signal should be like. On the Photon board five analog pins are available, which all generate PWM signals—these pins are labeled A0–A5.

Let's go back to our first experiment in this chapter where we flashed an LED on and off. We will use that same circuit, but instead of connecting the LED to digital pin D0, we can connect it to pin A0 as shown in the breadboard layout diagram in Figure 4.8.

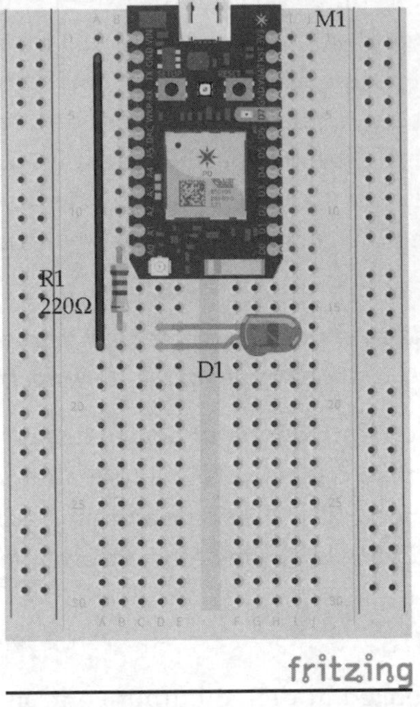

Figure 4.8 *LED connected to the analog pin on the Photon board.*

Because we are using the analog pin on the Photon board, we need to use a function called `analogWrite()` to generate the PWM signal and change the brightness of the LED. The analog pins on the Photon board are 8-bit values, which means you can write values from 0 to 255, as shown in the following:

```
int ledPin = A0;

void setup()  {
  // nothing happens in setup
  pinMode(ledPin, OUTPUT);
}

void loop()  {
  // fade in from min to max in increments of 5 points:
  for(int fadeValue = 0 ; fadeValue <= 255; fadeValue ++) {
    // sets the value (range from 0 to 255):
    analogWrite(ledPin, fadeValue);
    // wait for 30 milliseconds to see the dimming effect
    delay(30);
  }

  // fade out from max to min in increments of 5 points:
  for(int fadeValue = 255 ; fadeValue >= 0; fadeValue --) {
```

```
    // sets the value (range from 0 to 255):
    analogWrite(ledPin, fadeValue);
    // wait for 30 milliseconds to see the dimming effect
    delay(30);
  }
}
```

Run the program and see what happens. You should see the LED change from off to on, then on to off. This is because of the pattern in the `for loop`, which repeats itself indefinitely. In the first `for loop` i++ is shorthand code for i = i + 1, which always adds a plus one to the value of i. Similarly i-- subtracts 1 from the value; the first loop adds up the value to 255, and the second loop subtracts it back down to 0.

PWM signal control can be used in lots of circumstances to try and emulate a pure analog control over electronic devices. It is great for driving devices such as DC motors at varying speeds, but does not really work well for driving speakers, unless you drive them with some other circuitry components to smooth out the signal. Take a look at Figure 4.9 to see how the PWM signal looks at various stages.

PWM works by modulating the duty cycle of a square wave. The duty cycle refers to the percentage of time that a square wave is high versus low. The `analogWrite()` function sets the duty cycle of a square wave depending on the value that you pass to it:

- Writing a value of 0 with `analogWrite()` indicates that a square wave has a duty cycle of 0% which shows the LED to be off.
- Writing a value of 255 indicates that a square wave has a duty cycle of 100%, which would show the LED as always on.
- Writing a value of 127 indicates that a square wave has a duty cycle of 50%, which would show the LED to be dim.

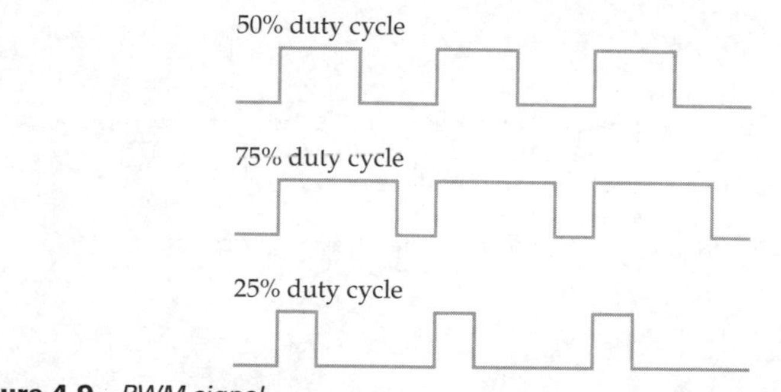

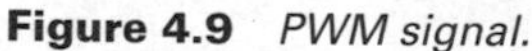
Figure 4.9 *PWM signal.*

Let's take a moment to think about what is actually happening here. We are not changing the voltage to a value between 0 and 3.3 V; we are simply switching from 0 V to 3.3 V at a very high frequency and according to how often (the duty cycle) we switch it. When you look at the LED, you don't even notice the switching because it is happening at such as fast rate that they eyes cannot process it fast enough, and the LED just seems either a bit dim or brighter than normal.

DAC

An alternative method to producing analog signals from the Photon board is to use the built-in DAC pins, which are pin A5 and the pin labeled DAC. The DAC produces an output voltage proportional to its current input, and it is very fast and accurate when doing so. This opens up more flexibility to your projects that regular PWM cannot do, such as audio output, as well as being able to filter out any digital noise that can occur.

Let's try the previous experiment where we dimmed an LED using PWM, but this time we will use a DAC pin on the Photon board. We might not see a noticeable difference in the two methods, but when the LED dims using PWM, sometimes you will notice some sort of flickering, especially when the value nearly reaches zero. With DAC, we shouldn't see any flickering at all and the brightness should be smooth. The values we can write to the DAC 12-bit output pin are between 0 and 4095 using the `analogWrite` function. The Photon board will check to see if it has a DAC pin and execute `HAL_DAC_Write` instead of the normal `HAL_PWM_Write`; this keeps the functionality and the coding simple to use.

```
int ledPin = DAC;

void setup()  {
  // nothing happens in setup
  pinMode(ledPin, OUTPUT);
}

void loop()  {
  // fade in from min to max in increments of 5 points:
  for(int fadeValue = 0 ; fadeValue <= 4095; fadeValue ++) {
    // sets the value (range from 0 to 4095):
    analogWrite(ledPin, fadeValue);
    // wait for 30 milliseconds to see the dimming effect
    delay(30);
  }
```

```
  // fade out from max to min in increments of 5 points:
  for(int fadeValue = 4095 ; fadeValue >= 0; fadeValue --) {
    // sets the value (range from 0 to 4095):
    analogWrite(ledPin, fadeValue);
    // wait for 30 milliseconds to see the dimming effect
    delay(30);
  }
}
```

In this example all we need to do is change the pin number to DAC, which is where we have connected our LED to the Photon board as shown in Figure 4.10. We also need to change the value, as we are working with a greater range than 0 to 255, which is now 0 to 4095. If you want to get a much smoother output, then you can also play around with the delay function by decreasing it until you get the effect you want.

You may be wondering what the point is in using PWM when we have DAC, which has a better output. Well, first, using PWM does not necessarily require any additional hardware and can be a very inexpensive alternative to using a natural analog signal. DAC devices usually require some sort of chip or additional hardware—some microcontrollers come with this built in and some do not.

Controlling a Servo Motor

DC motors are excellent drive motors, but they are not really good for precision work because there is no feedback. Without using some sort of encoder device you will never know the exact position of a DC motor. Servo motors,

Figure 4.10 *LED connected to the DAC pin on the Photon.*

or servos, are unique in that you can command them to rotate to a particular position and stay there until you tell it otherwise. A good example use of a servo motor is an actuating door lock.

When using servos, there are a couple of different types to consider, such as standard and continuous rotation. Standard servos operate from 0 to 180 degrees. Servo control is achieved by sending a pulse of a particular length. The length of the pulse determines the absolute position that the servo will rotate to. This is due to a small potentiometer in the servo that measures its position; when you remove the potentiometer, it becomes free to continuously rotate.

Unlike a standard DC motor, servo motors have three pins: power (red), ground (black or brown), and signal (white or orange). The wires are color coded and typically in order; they look like the ones shown in Figure 4.11.

Servos have a dedicated control pin, unlike standard DC motors, that tells the servo which position to turn to. The power and ground lines of a servo should always be connected to a power source. Servos are controlled using adjustable pulse widths on the signal line. For a standard servo sending a 1-ms 5-V pulse turns the motor to 0 degrees and sending a 2-ms 5-V pulse turns the motor to 180 degrees. Once a pulse has been sent to a servo, it then turns to that position and will remain there until instructed to move to another position by the sending of another pulse signal. If you want the servo to hold its position—that is, resist any movement, then you need to resend the same command every 20 ms or so.

In this experiment we are going to use a standard servo motor that operates from 0 to 180 degrees and we will control it using a simple potentiometer. We will read the value of the potentiometer, and as we rotate it from one

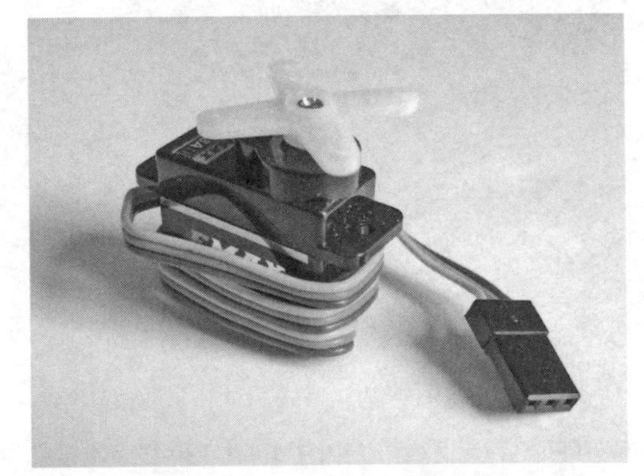

Figure 4.11 *Servo motor.*

value to another the servo motor will rotate accordingly. Because the value we read from the potentiometer and the value we want to write to the servo are very different, we can use a simple programming technique to scale down the value. The hardware we will be using for this experiment is shown in Table 4.6.

It's quite simple to connect the servo to the Photon board: connect the black or brown cable to any ground pin (GND), connect the red cable to the 5-V reference pin (VIN), and then finally connect the yellow cable to one of the analog pins, which will actually use PWM because we are sending a pulse to the servo. The potentiometer is wired in a similar way, where one of the pins is connected to ground, another is connected to 3.3 V, and the middle pin is usually connected to one of the analog input pins on the Photon so we can read its value. You can connect the potentiometer and the servo according to Figure 4.12.

Let's take a closer look at the code for controlling a servo:

```
int potPin = A0;
int servoPin = A1;
Servo myservo;

void setup()
{
  myservo.attach(servoPin);
}

void loop()
{
  int reading = analogRead(potPin);
  potPin = map(potPin, 0, 1023, 0, 180);
  myservo.write(potPin);
}
```

Schematic	Description	Appendix
M1	Photon board	M1
	Breadboard	H1
	Jumper wires	H2
	Servo motor	H5
	10-K potentiometer	R2

Table 4.6 *Hardware and Components*

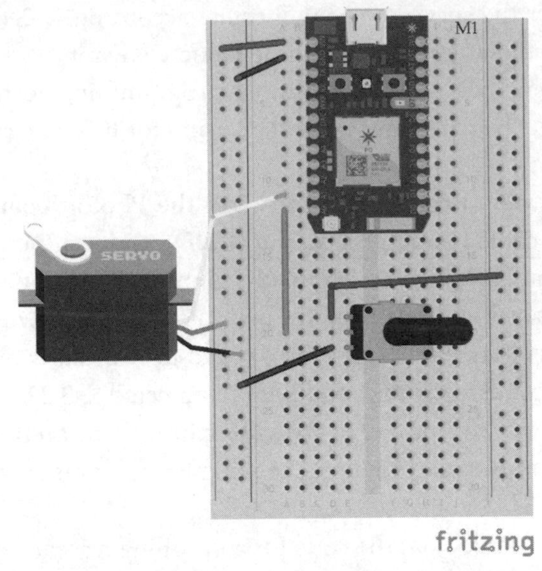

Figure 4.12 *Breadboard layout diagram for servo experiment.*

Normally, when using the Arduino platform you would have to include the "#include <servo.h>" library into the program, but by default it is already included behind the scenes, so this first line is unnecessary—the program will already look for the servo functions when we call them. The first thing we need to do is create an object in our program so whenever we tell the servo what to do, we will always call `myservo`. In the `setup()`, attaching the servo initializes everything necessary to control it. You can easily add multiple servos by creating different objects with different names and assign another analog pin to it. In the `loop()`, the potentiometer is read and scaled down using the `map()` function. The `map()` function takes one range of values and uses integer math to change it to another range of values. This is a useful function when you need to limit values.

```
Map(value, fromLow, fromHigh, toLow, toHigh);
```

Once the value has been read and scaled, it is then written to the servo to fix its position using a PWM pin to send the pulse. There is a short delay in the program just to make sure that the servo has reached its intended position before another command is sent to the servo again for a new position. You can see the final experiment in Figure 4.13.

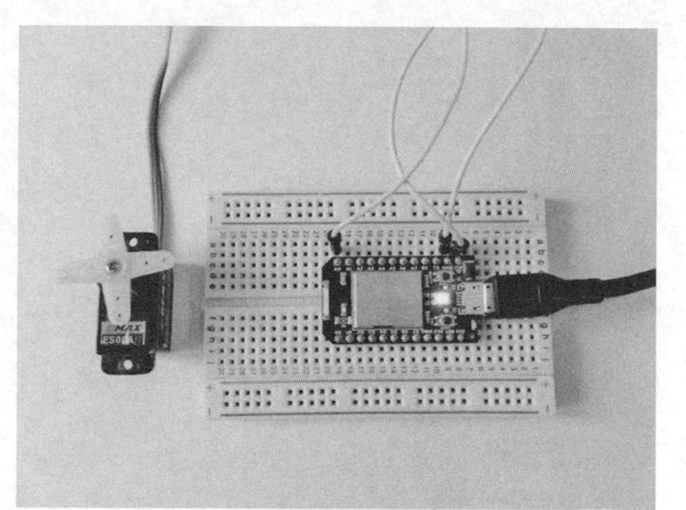

Figure 4.13 *Servo position.*

Summary

This concludes our chapter on digital and analog outputs for the Photon board. You should now have a good understanding of how both digital and analog pins work and what types of electrical components you can use to build your projects. We have also looked at some syntax for writing digital and analog values, as well as controlling a servo motor and creating and attaching an object. The next chapter is closely related to outputs, where we will control these outputs using some form of input device or electronic component.

5
Inputs

In this chapter we will program some input devices such as switches, temperature sensors, and many more. Input devices are generally used to trigger some output or event. A simple switch can easily be used to turn something on or off, such as a light-emitting diode (LED). A sensor used as an input device can be used to monitor something like temperature or a certain type of gas, and when triggered, it can set a series of events in motion or just be collected as data and interpreted into something visual. This chapter will also look at some of the basic programming used in reading and detecting inputs, which you can use for your own projects in the future. We will look at both digital and analog inputs, where a digital system will return a value of either 1 or 0 and an analog input will return a value from 0 to 4095 using some sort of analog-to-digital conversion. In this chapter you will accomplish the following:

- Understand how to read digital and analog inputs
- Experiment with using different types of input devices
- Learn more about coding with the Particle programming language

Digital Inputs

When we used `digitalWrite()` in the previous chapter, there were two different states that the LED could be: either HIGH or LOW. When it comes to using digital inputs, the same states apply to the switch—it is either HIGH or LOW. When you read the state of a digital input, it will be connected to either 3V3, which indicates a HIGH state, or to ground, which indicates a LOW state. By connecting a simple push-button switch to your Photon, you can easily change the state of a digital input pin by pressing it.

A simple push-button switch as shown in Figure 5.1 is great for experimenting with digital inputs, and is also an inexpensive component—something to consider when building your own circuits. You can see it snaps straight into the breadboard on either side of the gap. This switch is a common component you will become familiar with, as it is included in almost every electronic starter kit. When looking at the switch in Figure 5.1, the top two pins are connected together, as are the bottom two pins. When the push-button switch is pressed down, both sets of pins are connected in a circuit and it becomes complete.

You are going to use a small push-button switch to make or break a connection between 3V3 and one of the digital pins on the Photon, which you will configure in your firmware program that you will upload to the Photon. If you just connect the switch directly to the Photon board, this creates a problem when the switch is not closed, as the input pin is not connected to anything. This is referred to as floating, and could easily give you false readings when reading the state of the switch. When creating a switch circuit, you need

Figure 5.1 *A push-button switch.*

your readings to more precise, and the way to do this is using what is called a pull-down resistor.

Now think about what happens when the button is not pressed with the pull-down resistor in the circuit. The input pin will be connected through a 10-K resistor to ground. Although the resistor will restrict the flow of current, there is still enough to ensure that the input pin will read a LOW logic value. 10 K is a common pull-down resistor value. The value of the resistor must be low enough to make it resistant to any electrical interference, but also at the same time it must be high enough so that it prevents excessive current drain when the switch is closed. When the button is pressed, the input pin is directly connected to 3V3 through the button. In this circuit the current has two options:

- It can flow through a zero resistance path to 3V3.
- It can flow through a high-resistant path to ground.

When you connect the switch to your Photon, you need a pull-down resistor to ensure that the input is connected to ground on the Photon board. Table 5.1 shows the components that you will need to use in this example.

To wire up the circuit to the Photon board, insert your push-button switch into the breadboard as shown in Figure 5.1. If you find that it does not snap into the breadboard easily, then there is a possibility that the orientation is not correct—you may have rotate it 90 degrees. Using one of the jumper wires, connect digital pin 0 on the Photon to one of the top pins on the push button. Take another jumper wire and connect the bottom push-button pin to the 3V3 pin on the Photon. Using the 10-K resistor (brown, black, orange, and gold rings), connect the top terminal of the switch to ground; this will be our pull-down resistor. You can see this in Figure 5.2 and in the breadboard layout diagram in Figure 5.3.

Schematic Reference	Description	Appendix
M1	Photon board	M1
	400-point breadboard	H1
	Jumper wires (M-M)	H2
R1	10-K resistor	R3
S1	Push-button switch (tactile)	H6

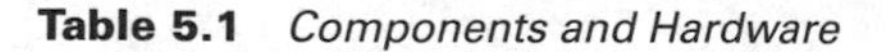

Table 5.1 *Components and Hardware*

Figure 5.2 *Connecting a push-button switch to the Photon.*

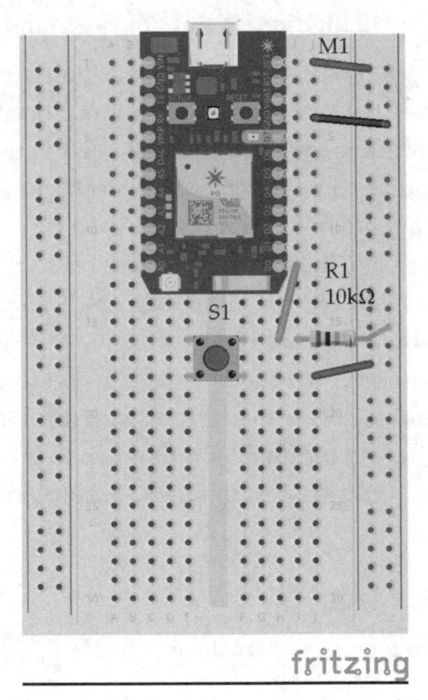

Figure 5.3 *Breadboard layout diagram for connecting a push button.*

Here is our basic digital input sketch, which will use the push button as the input device and light up the on-board LED, which is connected to digital pin 7 as an output:

```
int pushbutton = D0;
int led = D7;

void setup() {
    pinMode(pushbutton, INPUT);
    pinMode(led, OUTPUT);
}

void loop() {
    int buttonstate = digitalRead(pushbutton);
    if (buttonstate == HIGH){
            digitalWrite(led, HIGH);
    }
    else {
        digitalWrite(led, LOW);
    }
    delay(100);
}
```

We can look at the code in the following sections to make it easier for us to understand what it is doing:

- Set the push button (pin D0) as an output and the LED (pin D7) as an input.
- Read the state of the push button and store it in a variable called state.
- If the push-button state is HIGH or connected to 3V3, turn the LED on or HIGH.
- If the push-button state is LOW, turn the LED off or LOW.
- Pause the program for 100 milliseconds to slow down the output.

When you upload this program to the Photon and push the switch on your breadboard, you should see the on-board LED light up. When you release the push-button switch, the LED will switch off.

digitalRead ()

The main function that we can learn from our sketch is the use of `digitalRead()`. This checks the value of the pin, which is referenced in the parameters. In this example we refer to digital pin D0 and we call `pushbutton` to check if the push button is connected to 3V3 or to ground. `digitalRead()` returns a value of either HIGH or LOW, and we store this value in a variable called `state`.

You may also note that we used `pinMode()` to set the digital pin 0 as an input in the setup function. This is required so that the Photon knows how to treat that pin and allows us to use the function `digitalRead()` on that particular pin.

The syntax for `digitalRead()` is

```
digitalRead(pin);
```

The parameter of `digitalRead` is

pin, refers to the digital pin number
`digitalRead()` returns HIGH or LOW

Local and Global Variables

In the previous code examples the variables were declared outside both the `setup` and `loop` functions. Those variables are always referred to as *global variables* because they can easily be accessed and changed from either the `setup` or `loop` functions. In our example code we declared a new variable within the `loop` function:

```
int state = digitalRead(pushbutton);
```

When you call a variable within a block of code, it can only be accessed within that block. This is known as a *local variable* because it sits within a function. When the program is running on the Photon and it finishes executing a particular block of code, any local variables within that block are automatically freed up in memory so they can be used for other variables in the next block of code.

Debouncing

When you press down on a push button, you would expect to just get a single change from 1 to 0. However, in reality, that does not always occur. Sometimes when the contacts in the push button come together, they bounce when you release the button and create static signals. So now a single button press has now become two or more presses, depending on the push-button switch. All of this happens in a split second—the total amount of time the button registers its press is less than 200 milliseconds. Most new tactile-type switches may not bounce at all; however, a very old switch may have a lot of bounce. Sometimes the bouncing does not have any effect on the outcome of our sketch. For example, in our previous sketch we detected a push-button press, which then turned on an LED. Debouncing makes no difference to the outcome because when we release the switch, it will stabilize itself and the LED will switch off; this may take only a few milliseconds so we do not notice the debouncing effect.

One situation where debouncing may cause your outcome to be different from what is expected is if we use a push-button switch to turn an LED on or off every time the switch is pressed. When you press the button, the LED comes on and stays on; when you press the button again, the LED turns off and stays off. If you had a button that bounced every time you pressed it, then the LED would be on or off based on whether you had an odd or even number of bounces. Using the same circuit as our previous sketch, try flashing the following program to your Photon:

```
int ledpin = D7;
int ledValue = LOW;

void setup() {
pinMode (D0, INPUT);
pinMode (ledpin, OUTPUT);
}

void loop() {
    if (digitalRead(D0) == HIGH) {
        ledValue = ! ledValue;
        digitalWrite(ledpin, ledValue);
    }
}
```

Press the push button a few times and see what happens. You may notice that when you push the button that the LED may not turn on or off as it should, but when you push it a few more times, it does turn on or off. As previously explained, this is a perfect example of how debouncing occurs. Try loading the following sketch with the addition of the highlighted code in bold:

```
int ledpin = D0;
int ledValue = LOW;

void setup() {
pinMode (D1, INPUT);
```

```
pinMode (ledpin, OUTPUT);
}

void loop() {
    if (digitalRead(D1) == HIGH) {
        ledValue = ! ledValue;
        digitalWrite(ledpin, ledValue);
        delay (200);
    }
}
```

Press the push button again a few times—notice anything different? This time it works fine after inserting a short delay into the program. This is because as soon as the program registers the first push of the switch, it delays the program before it checks again just in case there is another bounce on the switch.

Sometimes when writing your code you may need to reverse a value from HIGH to LOW. You can do this with Boolean logic using the `!` or `not` operator:

```
ledValue = ! ledValue;
```

In your program you used this to reverse the value of the LED. You set the LED value as a global variable at the start, which was LOW, so the equation is "ledValue is equal to not LOW," which becomes HIGH. So when we use `digitalWrite` your value used to light up the LED is HIGH.

Analog Inputs

As you have learned, with digital inputs, the information you receive from a digital device is either HIGH or LOW, ON or OFF. In contrast to this there are a number of devices that can have a range of values, such as dials, sliders, temperature sensors, and many more. Analog inputs on the Photon give us a value ranging from 0 to 4095, depending on the voltage. The Photon has up to six analog pins as shown in Figure 5.4.

The analog pins on the Photon will accept any voltage between 0 and 3V3. As previously mentioned, because we are using a digital system, these values must be converted into digital using something called an analog-to-digital convertor (ADC), which is featured on the Photon board.

Luckily for us we do not have to understand how the ADC works in too much detail, as the software we will write will take care of most of it for us.

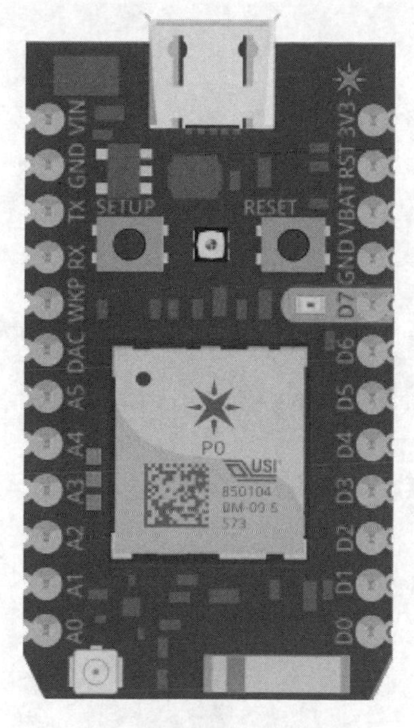

Figure 5.4 *Photon board analog pins.*

Reading a Potentiometer

The easiest analog sensor that we can read is a simple potentiometer (POT). You can find these in most electronic kits, and they are fairly common and inexpensive; you can even salvage these from old consumer electronics such as stereos, speakers, thermostats, and more. A potentiometer is a variable-voltage divider that looks like a dial knob, as shown in Figure 5.5. They can vary in size and shape, but they all have one thing in common, which is that they all use three pins. In our case, you connect one of the outer pins to ground and the other outer pin to 3V3. Potentiometers are also symmetrical, which means it does not matter how you connect the ground and 3V3, as long as it is not the middle pin. The middle pin connects to the analog pin on the Photon, as shown in Figures 5.5 and 5.6.

As you turn the potentiometer you vary the voltage that feeds directly into the analog pin on the Photon between 0 and 3V3. If you have a digital

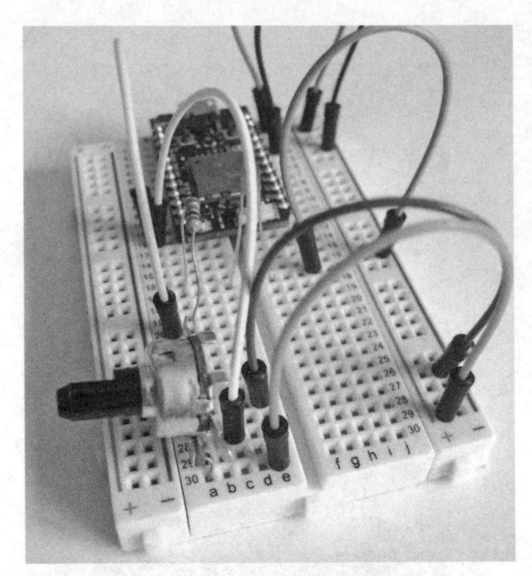

Figure 5.5 *Potentiometer connected to the Photon.*

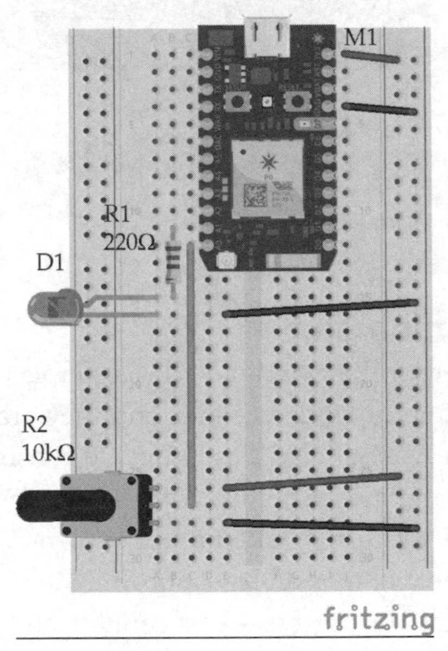

Figure 5.6 *Breadboard layout diagram for reading a potentiometer.*

multimeter handy, you can confirm the value when you turn the knob—simply change your multimeter value to voltage, hook up the red probe (positive) to the middle pin on the POT, and hook up the black probe (negative) to whichever side you have the POT connected to ground.

Here is the basic example sketch for reading analog input and using the value to create some output:

```
const int POT = A0;
const int led = A1;
int val = 0;

void setup() {
pinMode(led, OUTPUT);
pinMode(POT, INPUT);
}

void loop() {
int val = analogRead(POT);
int ledval = map(val, 0, 4095, 0, 255);

analogWrite(led, ledval);

delay(200);
}
```

We can take a look at the code in the following stages to understand better what it is doing:

- Create an integer called val and assign it a value of 0.
- Scale the analog read value, which runs from 0 to 4095, to the analog output value, which ranges from 0 to 255.
- Write the potentiometer value to the LED.
- Pause for a few milliseconds and repeat the process.

analogRead

Just like the function `digitalRead`, which returned a value of either HIGH or LOW, `analogRead` returns a value as well. In contrast to HIGH or LOW, `analogRead` returns an integer value from 0 to 4095. That value represents a voltage from 0 to 3V3.

The return value can be stored in a variable for use later on in our code, or it can be used right away using an `if` statement such as:

```
if (analogRead(POT) > 2000 {
      analogWrite(led, 129);
}
```

In our sketch the `analogRead` value is stored in a variable called `val`.

The syntax for `analogRead()` is

```
analogRead(pin);
```

The parameter is

Pin, the analog pin that the sensor is connected to on the Photon.

`analogRead()` returns an integer value between 0 and 4095, which represents ground and 3V3.

Const

In our code we introduced a couple of new programming concepts that we can use along the way. We previously used variables to store our pin numbers in memory; however, in our firmware we will use something called a constant:

```
const int POT A0;
```

This creates a spot of memory just like a variable does for an integer value called POT and stores the pin value A0. This seems like a variable but it isn't—when using a constant, you cannot change or edit the value that is assigned to it, as the name suggests. This can be useful when you want to store a value in the memory and you know that this value should never be changed within the code. If you do change your code, then you will receive an error when verifying your sketch.

map()

Because the returned value of `analogRead()` is a value between 0 and 4095, you may find that you need to scale this range of values down into something a bit more manageable. In our case what we want to do is receive the value from the potentiometer and then write that value to the LED using `analogWrite`. Because the `analogWrite` function has a range from 0 to 255, we cannot simply write the value of the POT because it would be too high. For this we can use a function called `map()` to scale down the values to something we can use. We can scale the input value using the following code:

```
int ledval = map(val, 0, 4095, 0, 255);
```

The `map()` function is responsible for scaling one set of ranges to another. The input scale we use is from 0 to 4095, and the output scale is from 0 to 255. If we calculated this scaling range manually, it would become too complicated and sometimes difficult to work out; it could also return recurring values of an infinite nature.

The syntax of `map()` is

```
map(input, inform, inTo, outFrom, outTo);
```

The parameters for `map()` are

- **input** the input value to be scaled
- **inform** the first number in the input scale
- **inTo** the second number in the input scale
- **outFrom** the first number in the output scale
- **outTo** the second number in the output scale

`map()` returns a value on the scale of `outFrom` to `outTo`.

You can easily modify the `map()` function by converting a value into a percentage from 0 to 100—just change the output values.

Variable Resistors

Most analog sensors work like a potentiometer does, calculating the varying voltage as the resistor changes its value. These are called variable resistors that resist the flow electricity through a circuit. A good example of this is a simple light-dependant resistor or photocell like in Figure 5.7. This changes the resistance based on the amount of light that hits it. When the light increases, the resistance goes down; therefore, the voltage in the circuit goes up. Take the light away, and the resistance increases and the voltage is reduced.

In order to read sensors like these with the Photon, you will need to create a voltage divider circuit and connect it to the analog input pins on the Photon board.

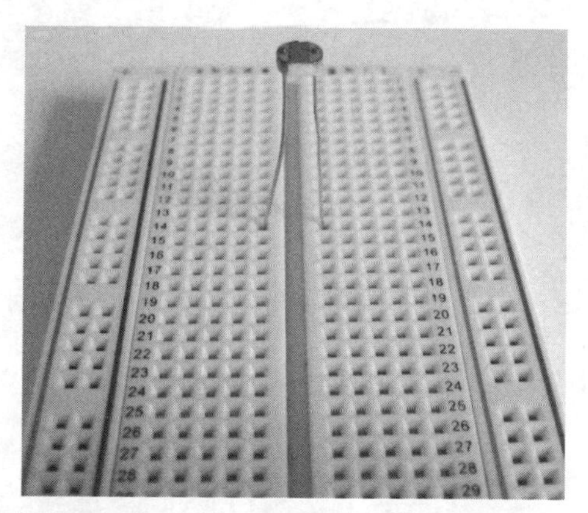

Figure 5.7 *A typical photocell, which acts as a variable resistor.*

Voltage Divider Circuits

When you are working with different sensors that offer a variable resistor feature, you need to create something called a voltage divider circuit. A voltage divider circuit converts the variable resistance into a variable voltage so we can read this value from the input pins on the Photon. In the schematic diagram in Figure 5.8 you can see a simple voltage divider circuit.

Figure 5.8 shows two resistors set up in series to one another between the input voltage and ground. You can also see one wire coming from between both resistors, which is the voltage output, which is the value that we read from the input on our Photon board. If we first consider a fixed voltage divider, we can understand the concept of how a voltage divider works. The mathematical calculation for working out the values of a voltage divider is as follows:

Vout = Vin(R2 / (R1 + R2))

In our case, the voltage input from the Photon would be 3V3 and the voltage output would be connected to one of the analog input pins on the Photon board. If we use the resistor values for R1 and R2 so that they are matched (both 10 K in this example), the 3V3 is divided by 2 to make an output voltage of 1.65 V according to the equation. Let's look at this in a bit more detail by adding our values to the following equation:

Vout = 3V3 / (10 K(10 K + 10 K)) = 3V3 * 0.5 = 1.65 V

Now what happens when we replace one of the resistors with a variable resistor such as a photocell? In this case we will replace resistor R1 with a

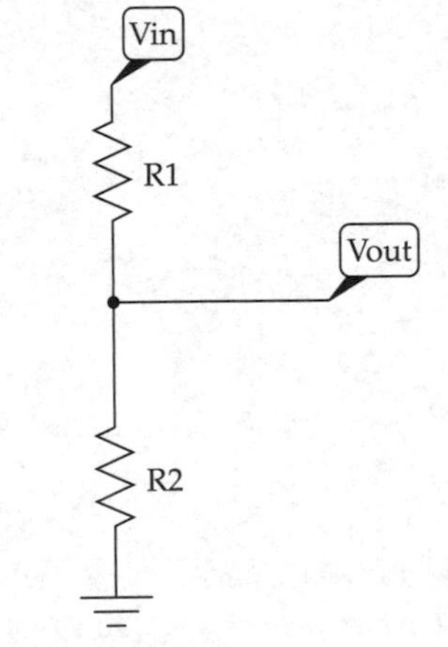

Figure 5.8 *Voltage divider circuit.*

200-K photocell. Whether you choose to replace R1 or R2 and which value you choose will affect the overall scale and precision of the output readings you receive. It's always worth experimenting with different configurations of resistor values until you find something that you are comfortable using and you are certain that the results are adequate.

For this example we are going to wire up a photocell and use it to determine the color of an RGB (red, green, blue) LED. Connect the circuit to your Photon as shown in Figure 5.9 and use the contents in Table 5.2.

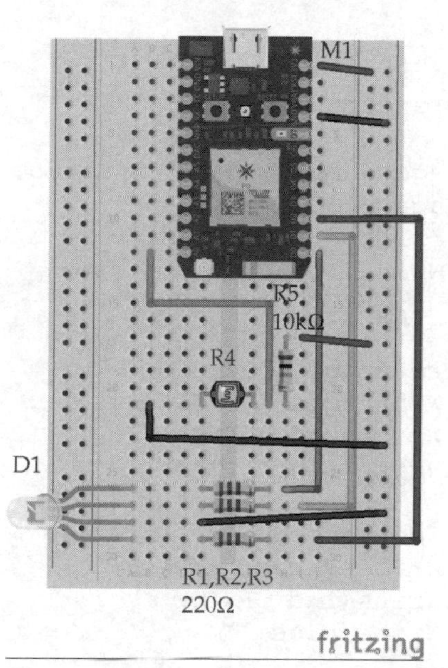

Figure 5.9 *Breadboard layout diagram for photocell voltage divider circuit.*

Schematic Reference	Description	Appendix
M1	Photon board	M1
	400-point breadboard	H1
R4	Photocell (200 K)	R4
R5	Resistor (10 K)	R3
	Jumper wires	H2
D1	RGB LED	S3
R1, 2, 3	220-ohm resistor	R1

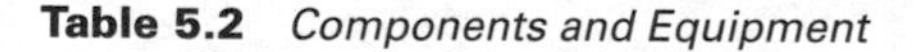

Table 5.2 *Components and Equipment*

Here is the sketch for reading the light levels from a photocell:

```
int red = D0;
int green = D2;
int blue = D1;

void setup() {
pinMode(red, OUTPUT);
pinMode(green, OUTPUT);
pinMode(blue, OUTPUT);

}

void loop() {
    int value = analogRead(A0);

    int percentage = map(value, 0, 4095, 0, 100);
    if (percentage < 33) {
        digitalWrite(green, HIGH);
        digitalWrite(red, LOW);
        digitalWrite(blue, LOW);
    }
    else if (percentage > 33 & < 66) {
        digitalWrite(green, LOW);
        digitalWrite(red, LOW);
        digitalWrite(blue, HIGH);
    }
    else if (percentage > 66) {
        digitalWrite(green, LOW);
        digitalWrite(red, HIGH);
        digitalWrite(blue, LOW);
    }
}
```

When you run the program on the Photon, the LED should light up a particular color depending on the light levels taken from the photocell. Try covering the photocell with your hand and see if the color changes again; then try shining something bright at the photocell, and it should change the LED color. Because we are using an RGB LED, we can use a number of different combinations of colors, so we are not necessarily forced to use three different colors—we can easily use up to six with the function `digitalWrite`. However, if we use an analog output, then the number of colors and shades we can create is limitless.

We can look at the code in a bit more detail by breaking it up into the following sections. When programming code, it is always easier to break your code up into sections, as it makes it easier to understand what's going on and easier to debug your code when there are issues.

- Declare the digital pins on the RGB LED. Label these with the color that each pin will represent.
- Tell the Photon that the RGB LED pins are digital output pins by using the function `digitalWrite`.
- Read the input value from the light-dependant resistor and store this as an integer in the variable called "value."
- Use the function `map()` to convert the input value to a percentage.
- Calculate the value using `if` statements to determine which color on the RGB LED to switch on and off.
- Add a short delay to the end of the code.

Within our firmware we use `if else` functions, which allows us to check another condition if the first condition is `false`. The syntax looks like this:

```
      if (conditions A) {
            execute the code here if A is true
      }
      else if (condition B) {
            execute the code here if condition A is
false and condition B is true
      }
      else {
execute the code here if both conditions A and B are
both false. This will always be the default option
}
```

You can use as many `if else` statements as you like within the code, or you may have an `else` statement at the end of the chain for the code that should be executed if all conditions return `false`.

Summary

In this chapter we have looked at some of the programming features that are used to control both analog and digital inputs on the Photon. These programming functions will be the key to creating your very own projects with the Photon. You should now be able to create your own circuits using both inputs and outputs.

In the next chapter we will look at controlling circuits using the Internet, from a simple command line to a Hypertext Markup Language (HTML) website.

6

The Internet of Things

Now we have learned the basics of programming the Photon board using various electronic components as well as using both analog and digital devices, we are now going to learn how to control those devices over the Particle cloud. The Internet of Things is gaining in importance with ever-increasing access to the Internet and networks. The Photon allows you to connect the board to the Internet using the built-in Wi-Fi chip, and this opens up endless possibilities with the board.

In this chapter we are going look at controlling things over the Internet as well as take readings from temperature devices to display on the Web. For this, we will be looking closely at using Particle functions, which is an important part of the Photon cloud.

Functions

Usually programming devices to access the Internet can be complex and time consuming. Luckily the Photon uses a pretty simple method of using functions within the program that either push or pull to the Web using a unique identifier for your Photon device. This first example we will look at creates a simple program to turn a light-emitting diode (LED) on or off using the Internet. This will help you understand how the Photon cloud system works, as well as introducing functions into your programming code.

A function command is quite simple—it associates itself with your Particle device and allows you to command it to do something. Whenever it receives a particular command, it then runs the script in your program. Sending commands to the Photon board actually requires you to send a Hypertext Transfer Protocol (HTTP) post request to the device. An easy way to test this is to use a simple command-line tool called *curl*. If you are using a Mac computer or a Linux device, chances are this tool is already installed in the operating system. Unfortunately, if you are using a Windows device, you will need to install it manually by following the next steps.

Open up your default Web browser and go to http://curl.haxx.se/download. Scroll down the page and find the Windows sections for downloads. Download the ZIP file and extract the contents to a new folder on your computer. Open up the command prompt by clicking Start and do a search for "cmd." This should open up the command prompt as shown in Figure 6.1.

Change the current directory to where you saved the contents of the ZIP file. You can do this by typing "cd /file location/" (refer to Figure 6.1 for clarification). Once in the directory we now should be able to run the command `curl`, which will list a few commands that we can use initially. The command `curl` is what we are going to use to send HTTP requests without having to use a Web browser. This method should be used first to test out the functions before you start building a webpage.

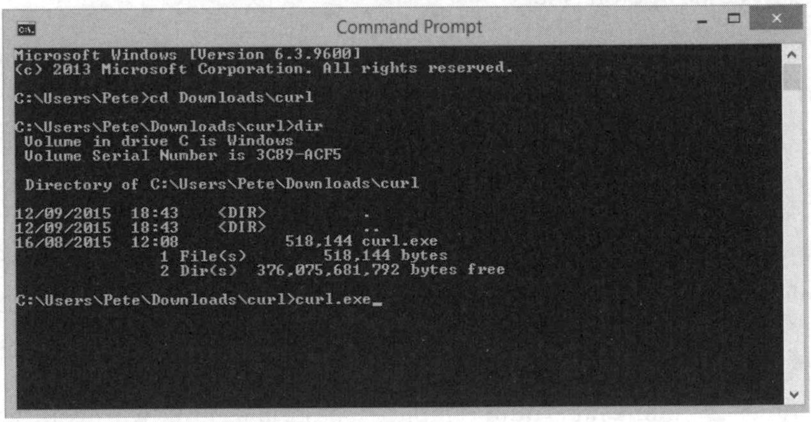

```
Command Prompt
Microsoft Windows [Version 6.3.9600]
(c) 2013 Microsoft Corporation. All rights reserved.

C:\Users\Pete>cd Downloads\curl

C:\Users\Pete\Downloads\curl>dir
 Volume in drive C is Windows
 Volume Serial Number is 3C89-ACF5

 Directory of C:\Users\Pete\Downloads\curl

12/09/2015  18:43    <DIR>          .
12/09/2015  18:43    <DIR>          ..
16/08/2015  12:08           518,144 curl.exe
               1 File(s)        518,144 bytes
               2 Dir(s)  376,075,681,792 bytes free

C:\Users\Pete\Downloads\curl>curl.exe_
```

Figure 6.1 *Windows command prompt.*

Schematic Reference	Description	Appendix
M1	Photon board	M1
	Breadboard	H1
	Jumper wires	H2
D1	5-mm LED	S1
R1	220-ohm resistor	R1

Table 6.1 *Components and Hardware*

Controlling an LED over the Internet

This experiment is similar to the first digital output experiment we did in Chapter 4. We are going to connect an LED to digital pin D0 and send a command over the Internet to turn the LED on and off. This is the basic principle of how to use the command, but in theory you can use it turn almost anything on or off. The hardware we will be using can be seen in Table 6.1 and the breadboard layout diagram in Figure 6.2.

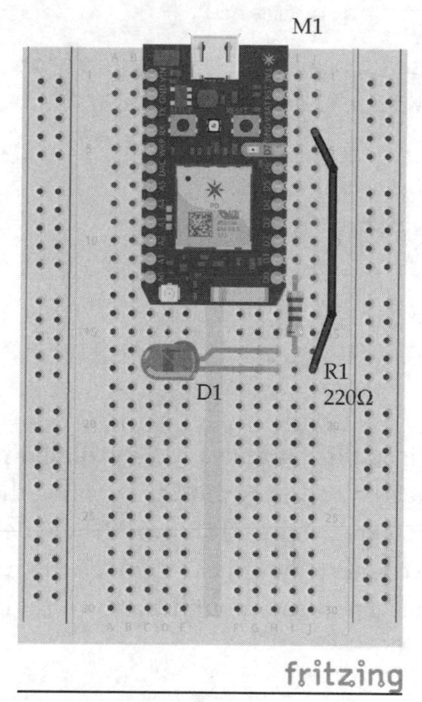

Figure 6.2 *Breadboard layout diagram for LED.*

Let's take a look at the following code, which we will use to turn the LED on and off using the Internet:

```
int led1 = D0;

void setup()
{

   pinMode(led1, OUTPUT);

   Spark.function("led",ledToggle);

   digitalWrite(led1, LOW);

}

void loop()
{

}

int ledToggle(String command) {

    if (command=="on") {
        digitalWrite(led1,HIGH);
        return 1;
    }
    else if (command=="off") {
        digitalWrite(led1,LOW);
        return 0;
    }
    else {
        return -1;
    }
}
```

The code first defines the variable `led1` for digital pin D0, so whenever we call `led1` we are actually calling pin D0. Labeling the pin makes it easier to know which pin you are using and allows efficient debugging. This is also the pin that we have connected our LED to so we can turn it on and off. The setup function defines `led1` as an output pin so the Photon board knows how to handle it when we send a command to the pin. The second line is where we declare our function that we can call through the Internet:

```
Spark.function("led",ledToggle);
```

This is the code that sets up the function, giving it the name "led." The second parameter is the name of the function in the out program to run when the `led` function is called over the Internet. Before we actually start turning the LED on or off, we need to make sure that the LED is first in an off state, using the following command to write the status to the LED:

```
digitalWrite(led1, LOW);
```

We keep the `loop` function part of the program empty because we have created a separate function to catch the `led` command and we do not need to keep listening for it in the `loop` function. When the `ledToggle` function is called as a result of receiving a message from the Internet, it receives a string as a parameter. In the HTTP request that we will send to the Photon board, we will make sure that we either turn the LED on or off using the correct parameter. If the function received a value that is either on or off, it will return 1 to indicate a successful command received. If the command sent is unsuccessful, then it will return a value of –1, indicating it has failed the checking process.

Something to consider also is the security process of actually sending commands to the Photon board. This is an important process, as you do not want your device open to anyone on the Internet to start sending ghost commands to your Photon board. Certain measures can take place when you send your commands to the Photon board. The Photon board requires up to two tokens to help secure your device.

The first token is your device's unique identifier—each of your devices will have this unique token number to help identify your device, especially when you are using several boards at the same time. You can find your device's ID by checking your device in the Particle build integrated development environment (IDE) and selecting your Particle device, as shown in Figure 6.3.

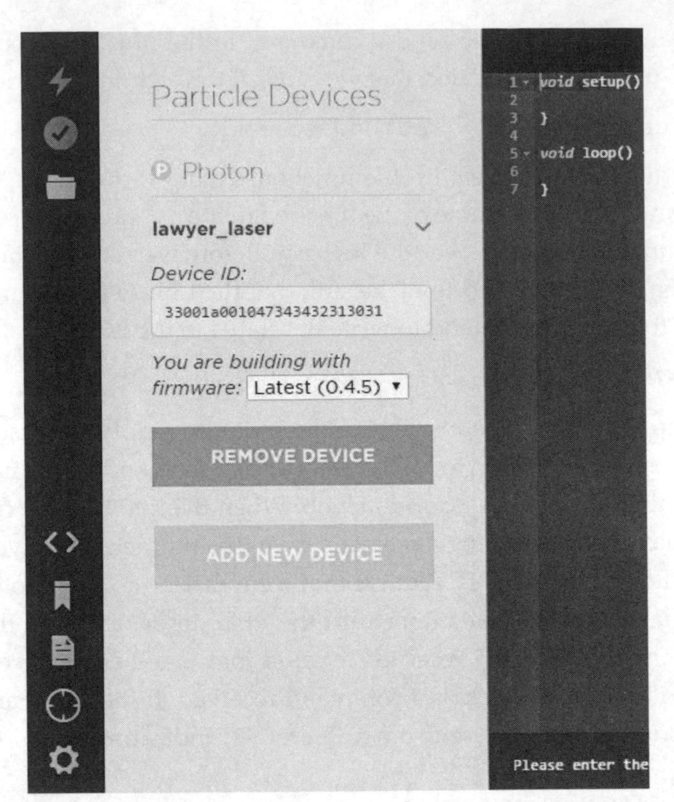

Figure 6.3 *Particle device ID.*

The second token that you will require is linked to your Particle account rather than the device. You can find this token ID from the Setting menu in the Particle IDE, where you will see your access token, as shown in Figure 6.4.

At any point you can generate a new unique access token if required; this is a good measure in case someone has accessed your token. You will need this token ID to send Web requests, so make a note of this token for later. You should now have something that looks like the following:

```
Device ID=55ff740626785011390716 67
Access Token=cb8b348000e9d0ea9e354990bbd39ccbfb57b30
```

Note that we have everything we need and the program is running on the Photon waiting for the `led` command to be sent. We can now send a

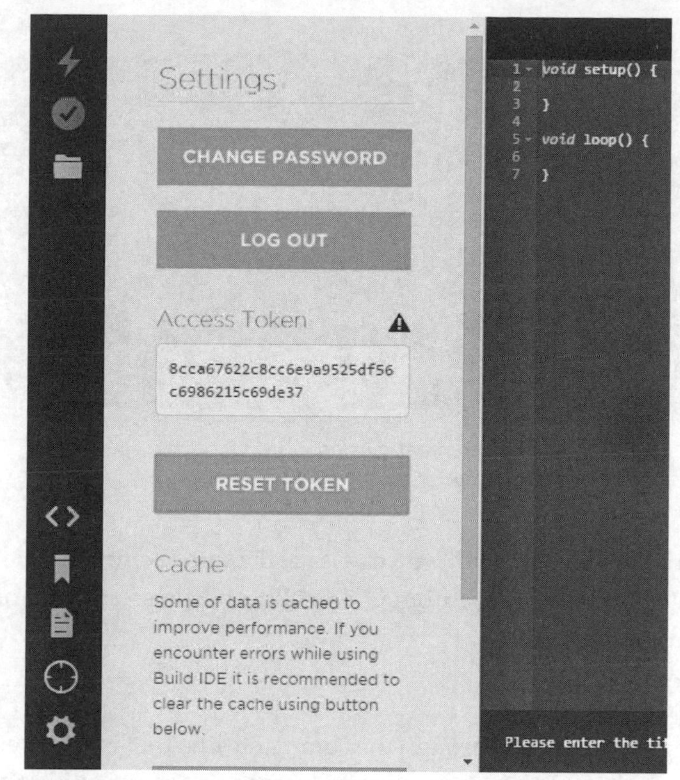

Figure 6.4 *Particle access token ID.*

command to it to turn the LED on using `curl` in the command line. To test this, we can send the following command:

```
curl https://api.spark.io/v1/devices/<deviceid>/led -d
access_token=<accesstoken> -d params=on
```

This command shows us how to use curl to send an HTTP request to control the LED. Before you do this, however, you need to change the device ID and access token with your token ID, which we discovered earlier. You can paste this command into the command line using `curl` as shown in Figure 6.5.

If everything went well, the LED connected to your Photon board should light up almost instantly, and you will see a response from your Photon device containing the information that it was successful by returning a value

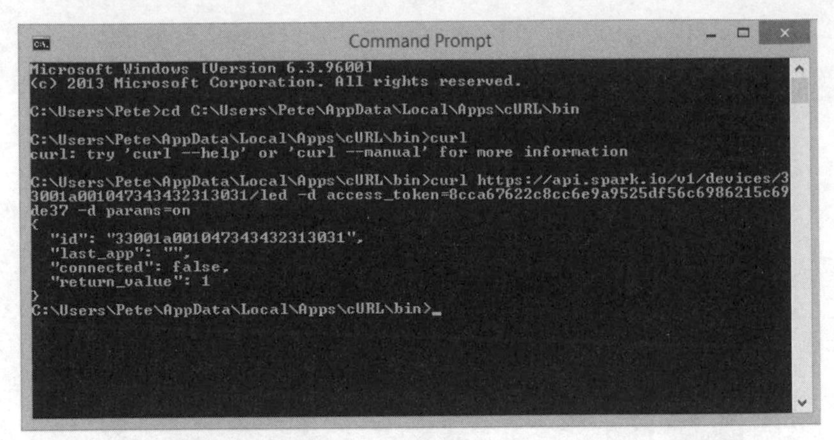

Figure 6.5 *Sending an HTTP command using* `curl`.

of 1. To turn the LED off again, you can issue the same command that we sent earlier, but change the value from ON to OFF and press ENTER. You should see the LED turn off.

Web User Interface

Controlling your devices through the command-line tool is an excellent way to test out your functions and circuits to make sure everything works as expected. But to get the best experience out of your projects, it would be much nicer to build a user interface using a webpage to control the Photon through the Web browser so that when we click a button, it will turn an LED on or off as shown in Figure 6.6.

When we register a function variable, we are basically making space for it on the Internet, similar to how you make space for a website that you would navigate to using your Web browser. We can create a simple Hypertext Markup Language (HTML) document on your computer with some basic buttons that send commands to the Photon board:

```
<center>
<br>
<br>
<br>
<form action="https://api.particle.io/v1/devices/your-device-ID-goes-
here/led?access_token=your-access-token-goes-here" method="POST">
Tell your device what to do!<br>
<br>
```

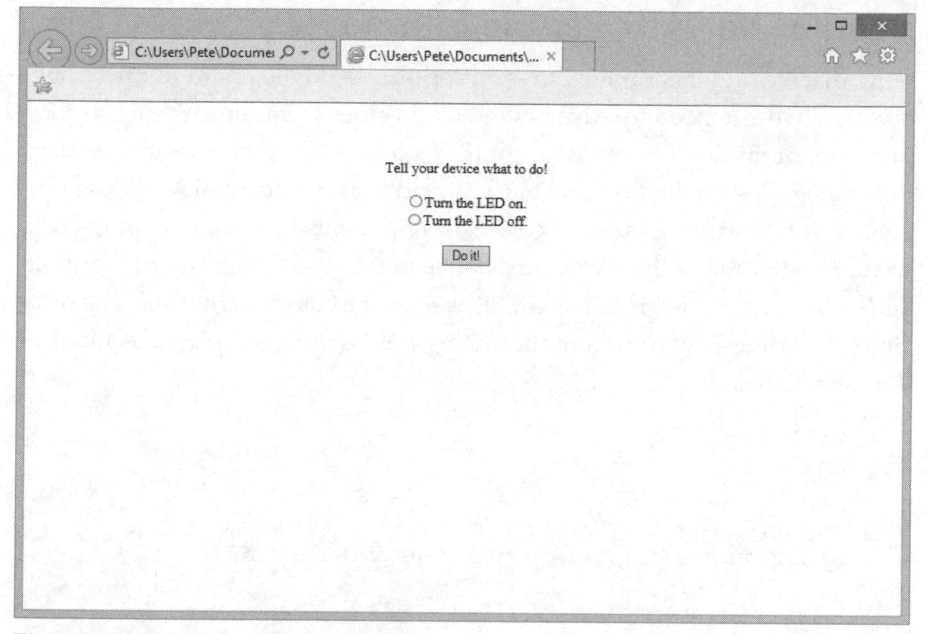

Figure 6.6 *LED Web browser interface.*

```
<input type="radio" name="args" value="on">Turn the LED on.
<br>
<input type="radio" name="args" value="off">Turn the LED off.
<br>
<br>
<input type="submit" value="Do it!">
</form>
</center>
```

Edit the code in the HTML document so that "your device id goes here" is your actual ID and "your access token" is your access token. You can open up a standard text editor to save this code as an .html document so that you can open it in your Web browser. Go ahead and open the .html document in your browser—you should be presented with a simple form that allows you to select either on or off. When you click the Do it! button you are posting information to the URL https://api.particle.io/v1/devices/your-device-ID-goes-here/led?access_token=your-access-token-goes-here. The information you are giving is the argument value ON or OFF. This parses through the `spark.function` that we registered. When you send the information, you will also get some information back from the page once sent that gives the status of your deice and lets you know that the post was successfully sent to the Photon board. If you want to go back, just click the back button in your browser.

Reading Values over the Internet

Now that we understand how to send commands to the board to turn things on or off, we also need to learn how to read values from sensors such as temperature, humidity, or light. We can refer back to Chapter 5 where we used the analog pins on the Photon board to read sensor information—these pins are labeled A0–A5. We can only connect components to the analog pin, where we know that the voltage can vary between 0 V and 3.3 V; any higher voltages will damage the Photon board as well as the component itself. The principle of reading anything from the analog pins is quite simple; take a look at the following code:

```
int getvalue = 0;
int analogPin = A0;

void setup() {
    Spark.variable("analog", &getvalue, INT);
    }

void loop() {
    getvalue = analogRead(analogPin);
    }
```

The program starts with a basic variable, which will be used to return a value when the webpage requests the function on the Photon board. The other variable is used to identify the analog pin that we are going to use to read from a sensor or similar device, in this case analog pin A0. In the setup function we link the two variables together using `spark.variable`. The first parameter is always the name given to the variable `(analog)`, and the second parameter identifies the variable that the sensor is connected to on the Photon board `(getvalue)`, and finally the last parameter defines the type of value, which is an integer value.

Upload the program to the Photon board to test it out. You can do this using the Web browser, because unlike functions, variables use HTTP GET requests, which can you can easily issue using the uniform resource locator (URL) in your browser. So with this in mind open up a browser and type the following URL into it:

```
https://api.spark.io/v1/devices/<device ID>/
analog?access_token=<Access token>
```

The request page should look like something Figure 6.7.

https://api.spark.io/v1/devices/33001a0
Apps Amazon Cloud Player Google Amazon.co.uk – Onli...

```
{
  "cmd": "VarReturn",
  "name": "analog",
  "result": 1295,
  "coreInfo": {
    "last_app": "",
    "last_heard": "2015-09-12T19:27:33.142Z",
    "connected": true,
    "last_handshake_at": "2015-09-12T19:26:54.599Z",
    "deviceID": "33001a001047343432313031",
    "product_id": 6
  }
}
```

Figure 6.7 *URL request in browser.*

You can reload the page a few times, which will resend the URL request to the Photon, and you should see a different value. Currently, as there is nothing connecting to the analog pin, the value is *floating*.

Reading a Light Sensor

In this experiment we are going to be using a simple photoresistor to measure light levels and display that information on a webpage with a nice interface that should look similar to Figure 6.8.

For this project we are going to be using a photoresistor, which is a light-dependent resistor that decreases its resistance value when light intensity increases and vice versa. Table 6.2 shows the hardware and components that we will be using for this experiment.

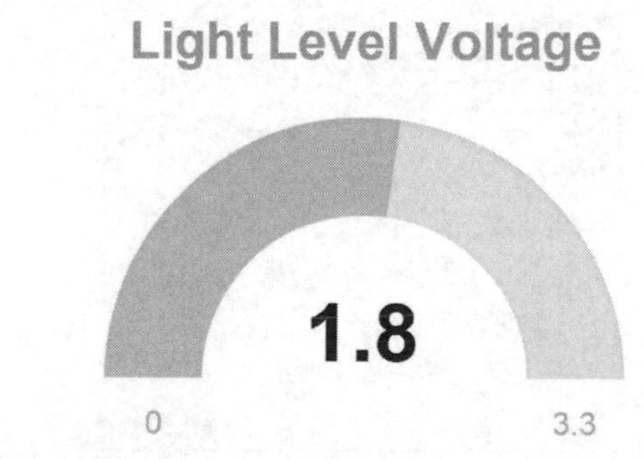

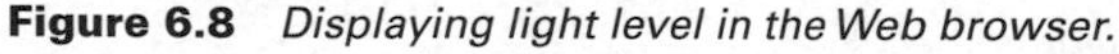

Figure 6.8 *Displaying light level in the Web browser.*

Schematic Diagram	Description	Appendix
M1	Photon board	M1
	Breadboard	H1
	Jumper wires	H2
R1	Photoresistor	R4
R2	10-K resistor	R3

Table 6.2 *Hardware and Components*

Figure 6.9 shows the breadboard layout diagram for this experiment.

Both the resistors and the photoresistor can be connected in the circuit either way round, and both are also nonpolarized. The resistors in a circuit like this create something called a voltage divider circuit. This is a really inexpensive way of taking an input voltage and by using both values and some mathematics, you get a desired voltage output. In our circuit we are using the photoresistor as a variable resistor to output a voltage that we can read on pin A0. When the light level changes, it also changes the resistance, thus changing the voltage. Once you have created the circuit as shown in the breadboard layout diagram, power up the Photon board so it is ready to load our program.

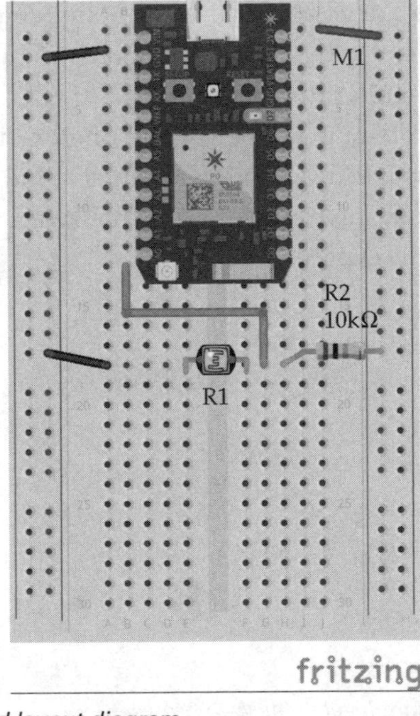

Figure 6.9 *Breadboard layout diagram.*

The program code we are going to use for this experiment is as follows:

```
int reading = 0;
double volts = 0.0;
int analogPin = A0;

void setup() {
    Spark.variable("analog", &reading, INT);
    Spark.variable("volts", &volts, DOUBLE);
    }

void loop() {
    reading = analogRead(analogPin);
    volts = reading * 3.3 / 4096.0;
    }
```

Looking at the code, we can see it is similar to the previous experiment's code, with the exception that we are now going to use two variables in our program. The first variable is the same as in the previous example and returns the same value when reading analog pin A0. The second variable will be used to return the actual voltage value at analog pin A0. There is a simple mathematical equation that we can used in the `loop` function to calculate the voltage. The value from the reading is multiplied by 3.3 V and then is divided by 4,095, which is the maximum value that can be read on the analog pin.

You may remember we used a bit of HTML to turn an LED on or off. The biggest issue with this is that the code can been seen in plain text by a user and, as such, causes some security concerns for those who want to host the webpage on the Internet. Using HTML on a local server or your own desktop computer is fine if you are testing out the functions and you are the only person able to access those files. However, in order to be safer and more secure, we can use a programming script language within HTML that hides all our code—more importantly, it hides our token and device IDs. We have also added a nice little graphical user interface, which gives a nice feel to the data that is being presented; this is done using a simple JavaScript plugin courtesy of www.justgauge.com. The HTML code for this experiment is as follows:

```
<html>
<head>
<script src="http://ajax.googleapis.com/ajax/libs/jquery/1.7.2/jquery.
min.js" type="text/javascript" charset="utf-8"></script>
<script src="raphael.2.1.0.min.js"></script>
<script src="justgage.1.0.1.min.js"></script>
```

```
<script>
var accessToken = "your access token here";
var deviceID = "your device id here"
var url = "https://api.spark.io/v1/devices/" + deviceID + "/volts";
function callback(data, status){
        if (status == "success") {
                volts = parseFloat(data.result);
                volts = volts.toFixed(2);
                g.refresh(volts);
                setTimeout(getReading, 1000);
        }
        else {
                alert("uh oh!");
        }
}
function getReading(){
        $.get(url, {access_token: accessToken}, callback);
}
</script>
</head>

<body>
<div id="gauge" class="200x160px"></div>

<script>
var g = new JustGage({
    id: "gauge",
    value: 0,
    min: 0,
    max: 3.3,
    title: "Voltage",
    levelColors: [
        "#00FF00",
        "#FFFF00",
        "#FF0000"
        ]
});
getReading();
</script>

</body>
</html>
```

The gauge JavaScript library is imported from the local .js files, which you can download from the www.justgauge.com website, and the variables are set up for both the access token and device ID—make sure that you change these to match the tokens for your device. The JavaScript function `getReading` is called to tell the HTTP request to be sent to the Photon board and then attaches the `callback` function. When the HTTP request responds, it checks to make sure that the request sent to the Photo was successful, and if it was then it retrieves the voltage value. The gauge display is then updated with the new voltage reading from the Photon by calling `g.refresh`. In the Div ID gauge you can determine many factors of the gauge such as color, min/max values, and labels. The current script reads the value of the light sensor every second; it may be worth changing this value to something a bit more appropriate, such as every 30 seconds or one minute, as updating the reading every second would be inefficient.

Open up the HTML page in your Web browser and you should see the current voltage reading. Cover the photoresistor with your hand and you should see that the value changes as it drops off. If you hold the photoresistor up to the light or flash a light of some kind into it, the voltage rating will go up.

This is the basic principle of reading a sensor with a resistance value. There are also other kinds of sensors that can be used in place of the photoresistor, such as a gas sensor or even a photoresistor dial.

In the next experiment we will look at reading a temperature sensor using a Maxim DS18b20 digital chip.

Reading a Temperature Sensor

For this experiment we are going to use what we learned in the previous project and use it to read a simple Maxim one-wire digital temperature sensor. We are going to use the same graphical interface for the gauge but change the parameters to display the temperature rather than the voltage. Figure 6.10 shows the webpage temperature reading from the digital temperature sensor.

This experiment uses the Maxim DS18b20 digital temperature sensor, which comes in all shapes and sizes and is more commonly found as an integrated circuit (IC). This digital temperature sensor can read temperatures in the range of –55° C to 125° C. These sensors are more commonly found in

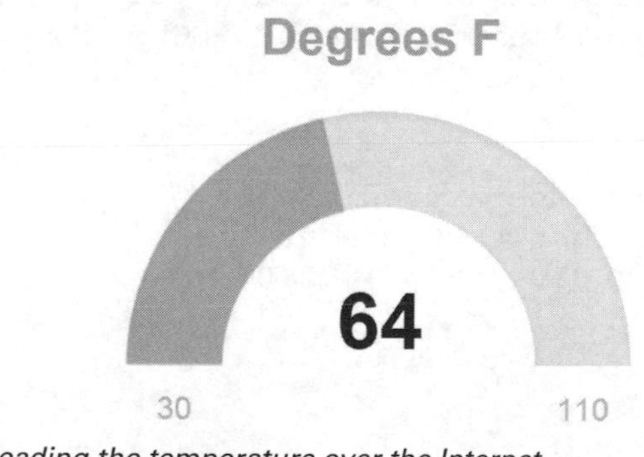

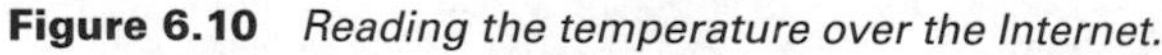
Figure 6.10 *Reading the temperature over the Internet.*

Schematic Reference	Description	Appendix
M1	Photon board	M1
	Breadboard	H1
	Jumper wires	H2
S1	Maxim DS18B20 IC	S2
R1	4.7-K resistor	R5

Table 6.3 *Components and Hardware*

thermostatic controls, industrial systems, or any other thermally sensitive system. The components that we are going to be using for this experiment can be found in Table 6.3.

Create the circuit using the breadboard layout diagram shown in Figure 6.11.

The resistor is used as a pull-down resistor to make sure there is no ghost reading from the sensor and gives us a more accurate reading.

Let's take a look at the program code that we will be using to read the temperature sensor:

```
#include "OneWire/OneWire.h"
#include "spark-dallas-temperature/spark-dallas-
temperature.h"

double tempC = 0.0;
double tempF = 0.0;

int tempSensorPin = D0;

OneWire oneWire(tempSensorPin);
DallasTemperature sensors(&oneWire);

void setup() {
    sensors.begin();
    Spark.variable("tempc", &tempC, DOUBLE);
    Spark.variable("tempf", &tempF, DOUBLE);
}

void loop() {
  sensors.requestTemperatures();
  tempC = sensors.getTempCByIndex(0);
  tempF = tempC * 9.0 / 5.0 + 32.0;
}
```

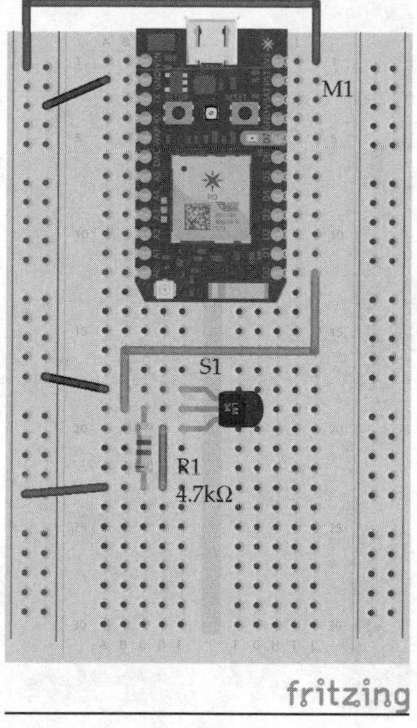

Figure 6.11 *Breadboard layout diagram.*

The first thing you will notice is that the program uses two different libraries, indicated by the `#include` statement at the top of the program. As we know from previous experiments, libraries are used so that we don't have to make our programming code really complex, and this allows us to include functions from other programs without having to understand the complexity. Because we are using a DS18B20 IC which uses a one-wire serial communication, we need to import the `OneWire` library, which will handle all the communication with the digital temperature sensor. The `spark-dallas-temperature` library handles everything else to do with reading the temperature on the Photon board—luckily someone has already created this for us. We add these libraries in the same way as before by using the build IDE in your Web browser and navigating to the Libraries section as shown in Figure 6.12. You can search for "spark dallas temperature" and add this to your application.

After we define the digital pin for the sensor, we must tell the program to start the one-wire serial bus on that particular pin. The next line directly

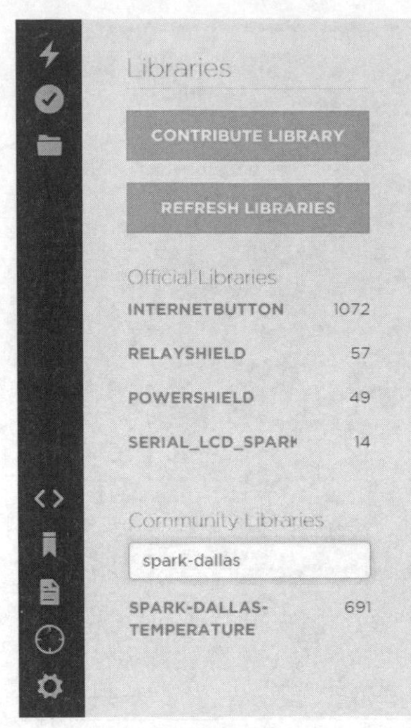

Figure 6.12 *Searching for the spark-dallas-temperature library.*

below that tells the library to use that particular interface when communicating serial data to the temperature sensor. Because of its nature, you can connect multiple sensors to the same pin.

```
OneWire oneWire(tempSensorPin);
DallasTemperature sensors(&oneWire);
```

The `setup` function defines the spark variables to read temperatures in both Fahrenheit and Celsius, which means we can display both on the Web. We must also call `sensors.begin()` to start the monitoring process. The `loop` function is used to request the reading and convert the values using some basic mathematics into our desired temperature readings. The first thing that we call is `sensors.requestTemperatures()`, which gets the reading from the temperature sensor. The second part accesses the reading itself using `sensors.getTempCByIndex(0)` and stores it in a variable called `tempc`; the number after the index is the sensor number, where 0 is the first, 1 is the second, etc. This only applies if you are using multiple sensors at the same time using the `OneWire` serial communication. Once we have the value in Celsius, we can then convert this to Fahrenheit using a standard formula.

The webpage is similar to the light sensor we created before, with a few minor changes to the attributes that are displayed:

```
var g = new JustGage({
    id: "gauge",
    value: 0,
    min: 30,
    max: 100,
    title: "Degrees F"
});
getReading();
```

We need to change the gauge range to match those of the temperature sensor, so the minimum value in degree F would be 30 and the maximum value is 100. If we want the temperature to be displayed in degrees C rather than degrees F, we need to change the following lines:

```
var url = "https://api.spark.io/v1/devices/" + deviceID + "/tempf";
```

to

```
var url = "https://api.spark.io/v1/devices/" + deviceID + "/tempc";
```

Open up the HTML page in your browser, and you should be able to read the temperature value from the sensor on the Photon as shown in Figure 6.13.

Measuring Distance Using an HC-SR04 Sensor

This experiment includes the use of an HC-SR04 ultrasonic range finder to detect the distance from certain objects, as well as being used as a motion detector to determine when something gets too close. These sensors are more commonly used in robotics to stop the robot from bashing into objects.

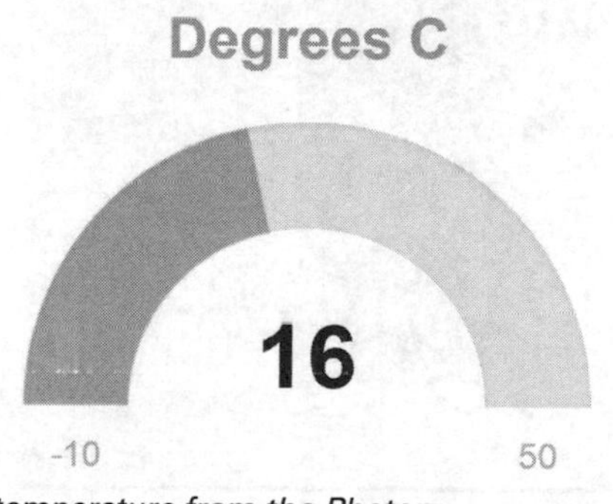

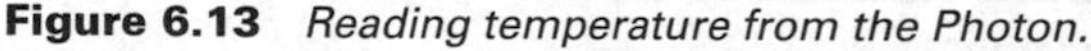

Figure 6.13 *Reading temperature from the Photon.*

Schematic Reference	Description	Appendix
M1	Photon board	M1
	Breadboard	H1
	Jumper wires	H2
H1	HC-SR04	M2

Table 6.4 *Components and Hardware*

The ultrasonic sensor sends out pulses every so often using one sensor, and the other sensor reads the feedback of those pulses and determines, based on the timings in between, how far away an object is. This works on the same principle as echolocation, which is how bats are able to see at night or how dolphins can detect their surroundings. For this experiment we are going to use the components listed in Table 6.4.

The ultrasonic range finder used for this experiment has four pins. Two of the pins are for 5-V power and ground; the other two pins are labeled trigger and echo. When the trigger pin is activated for a very short period, it sends out pulses of ultrasound. When the pulse is returned to the range finder the echo pin will indicate this. The breadboard layout diagram for this experiment can be shown in Figure 6.14.

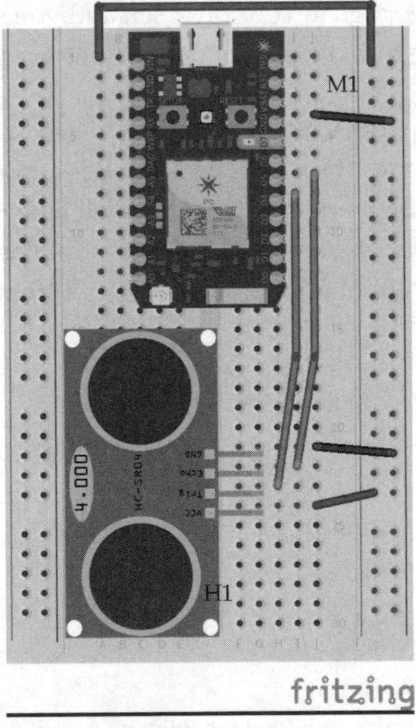

Figure 6.14 *Breadboard layout diagram.*

Let's take a look at the program code that we will use on our Photon board:

```
#include "HC_SR04/HC_SR04.h"

double cm = 0.0;
double inches = 0.0;

int trigPin = D4;
int echoPin = D5;

HC_SR04 rangefinder = HC_SR04(trigPin, echoPin);

void setup()
{
    Spark.variable("cm", &cm, DOUBLE);
    Spark.variable("inches", &inches, DOUBLE);
}

void loop()
{
    cm = rangefinder.getDistanceCM();
    inches = rangefinder.getDistanceInch();
    delay(100);
}
```

In the first line we can import the `HC_SR04` library so we can use all the functions to get the distance from the sensor. The library can be found in the Particle build IDE by searching for "HC SR04" and importing it into your program. For this experiment we are going to provide two options to measure in both centimeters and inches, which we define as double variables.

```
double cm = 0.0;
double inches = 0.0;
```

We also have created two `spark` variables for both centimeters and inches so we can call the values from our webpage for whichever one we require. The `loop` function keeps checking the distance every tenth of a second to update the value accurately. Our HTML page will be similar to the previous experiment, with a few minor changes:

```
<html>
<head>
<script src="http://ajax.googleapis.com/ajax/libs/
jquery/1.7.2/jquery.min.js" type="text/javascript"
charset="utf-8"></script>
```

```
<script src="raphael.2.1.0.min.js"></script>
<script src="justgage.1.0.1.min.js"></script>


<script>
var accessToken = "you access token here";
var deviceID = "you device id here"

var distance_url = "https://api.spark.io/v1/devices/" +
deviceID + "/distanceCM";

function callbackDistance(data, status){
      if (status == "success") {
            dist = parseFloat(data.result);
            dist = dist.toFixed(2);
            dist_gauge.refresh(dist);
            setTimeout(getDistanceReading, 1000);
      }
}

function getDistanceReading(){
      $.get(distance_url, {access_token: accessToken},
callbackDistance);
}
</script>
</head>

<body>
<div id="distanceGauge" ></div>

<script>
var dist_gauge = new JustGage({
    id: "distanceGauge",
    value: 0,
    min: 5,
    max: 250,
    title: "Rangefinder (cm)"
});
getDistanceReading();
</script>

</body>
</html>
```

Figure 6.15 *Range finder webpage.*

In the HTML all we really need to do here is change the values to match the variable in our program on the Photon. For this example we are going to measure the distance in centimeters, so we need to change the following URL to reflect that:

```
var distance_url = "https://api.spark.io/v1/devices/" +
deviceID + "/distanceCM";
```

You also need to change the script for the gauge to change the maximum and minimum values to those our range finder can detect. According to the datasheet, this is in the range of 5 to 250 cm. When you load the webpage in the browser as shown in Figure 6.15, you should see the distance of the nearest object to the sensor. Try moving your hand closer and refreshing the page; this should change the value.

Summary

In this chapter we attached digital LED outputs and analog sensors to the Photon board. We learned how to turn the digital outputs on and off using both the command-line interface and the Web browser using some basic HTML programming language. We created a nice visual element to display sensor information from light, temperature, and a range finder. You should now be able to apply these programs to any given device and hopefully control and create your very own electronic projects over the Internet.

In the next chapter we will be looking at some add-on boards for the Photon called shields. These will greatly enhance your project abilities without the need for creating complex circuits.

7

Programming Particle Shields

In this chapter we are going to be looking at Particle shields and add-on boards that are available and how these can make your projects a little bit easier without all the unnecessary circuit design and testing. Many shields are available that can assist with your projects, including power, relay, Joint Test Action Group (JTAG), Arduino shield, big button, and many more. We are going to be looking at all of these in detail and show you how you can use them in your projects.

Shield Shield

Sometimes when using two different electronic devices, the two do not match up because of the different voltages the two systems use. The Photon uses 3.3-V standard on the digital and analog pins, but more conventional devices are usually affixed with 5 V, and there is a significant risk of damaging both boards. The shield in Figure 7.1 performs all the necessary voltage translating while also providing a useful Arduino-compatible footprint to make it easier to plug in your existing Arduino shields or program other 5-V devices.

The shield is based on the Texas Instruments integrated circuit (IC) TXB-0108PWR, which handles all the voltage translation between the Particle Photon's 3.3-V to 5-V logic. This is only the case for the digital pins, not the analog pins, which are still rated at 3.3 V.

NOTE *Do not exceed 3.3 V at any time, or you will risk damaging the board.*

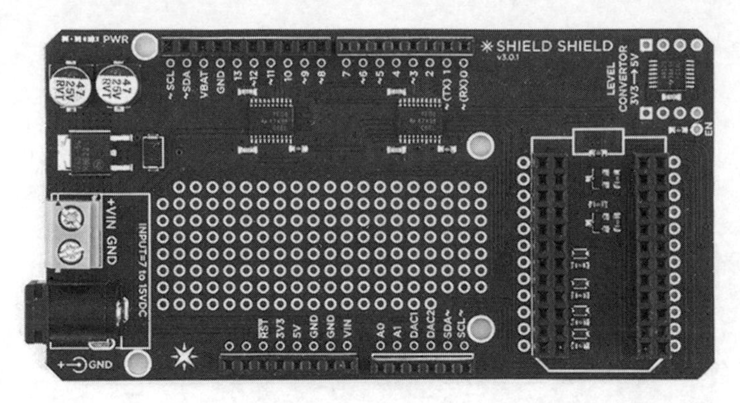

Figure 7.1 *Shield Shield.*

The shield also has an on-board voltage regulator and can be easily powered from a 7- to 15-V direct current (DC) power source. You can still power the shield from the universal serial bus (USB) port using the Photon board, but the current would be limited to 500 mAh.

You may have noticed all the dots in the middle of the Shield Shield board, which is a nice prototyping area for some small projects. This is a useful area when you want to test certain components on a regular basis. It should also be noted that one of the drawbacks of the Shield Shield is that it is only capable of driving loads at short distances, so with that in mind, longer wires will introduce some capacitance loading and may cause auto-direction detection to fail and as such will not work. There is, however, an on-board 74ABT125 buffer that is capable of driving heavier loads in one particular direction—you can use a simple jumper wire to whichever input/output (IO) pin you want to covert to 5 V. The pin mapping for the Shield Shield can be found in Figure 7.2, and the pin labels are shown in Table 7.1.

Shield	Photon	Peripherals
0	RX	Serial1 RX, PWM
1	TX	Serial1 TX, PWM
2	A2	SPI1_SS
3	WKP	PWM, ADC
4	D6	
5	D0	SDA, PWM
6	D1	SCL, PWM, CAN_TX
7	D7	
8	A5	SPI1_MOSI, PWM

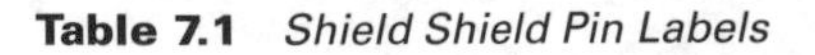

Table 7.1 *Shield Shield Pin Labels*

Shield	Photon	Peripherals
9	A4	SPI1_MISO, PWM
10	D5	SPI3_SS
11	D2	SPI3_MOSI, PWM, CAN_RX
12	D3	SPI3_MISO, PWM
13	D4	SPI3_SCK
A0	A0	ADC
A1	A1	ADC
A2	DAC1	DAC, ADC
A3	DAC2	SPI1_SCL, DAC, ADC
A4	D0	SDA, PWM
A5	D1	SCL, PWM, CAN_TX

Table 7.1 *Shield Shield Pin Labels (Continued)*

You can see in Table 7.1 that the Shield Shield does not map the Particle's pins to the Arduino's pins exactly; in other words, pin D0 on the Photon does not match the same pin D0 on the Arduino shield.

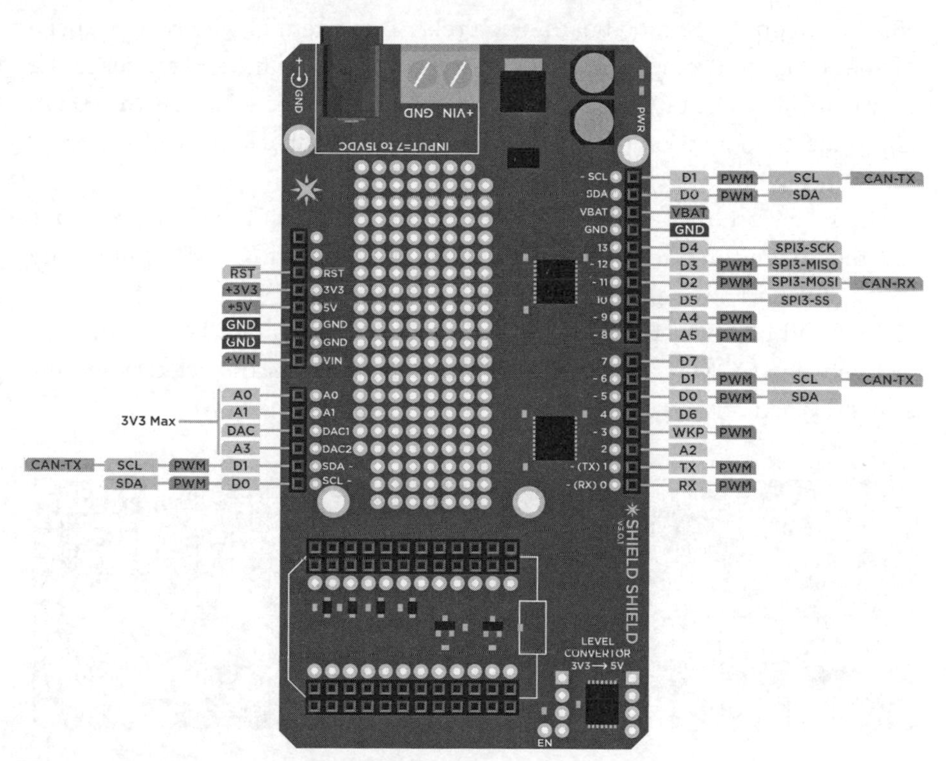

Figure 7.2 *Pin mapping for the Shield Shield.*

Relay Shield

The relay shield is just that—it allows you to control any electrical device using the Photon board. You can turn electrical appliances on and off, controlling them over the Internet, such as lamps, coffee machines, or anything that has a high voltage rating. The relay shield comes with four switches that are rated at a maximum of 220 V at 10 A, allowing you to control almost any electrical device under 2000 watts. You are not necessarily restricted to using only appliances—there are also other applications you can use the relays for that have a high voltage rating. You can see in Figure 7.3 the layout of the relay shield, and you may notice there is also an area on one side of the board for prototyping some small electronic components or connectors such as a temperature sensor or light sensor that could switch the light on and off.

The relay shield provides regulated power to the 5-V rails on the Particle device, as well as 5-V power to control the relay switching, but does not support power to any devices controlled by the relays. The use of the relay shield is actually quite simple—it has four relays that are controlled by pins D3, D4, D5, and D6 on the Photon board. Each relay is triggered by an NPN transistor that takes control of the signal from the Photon board and switches the relay coil either ON or OFF. There is also a diode connected across the coil to help protect the transistor from any high-voltage feedback, which may occur due to the switching.

The relays are the single-pole, double-throw (SPDT) type, which means that they have three terminals at the output: common (COMM), normally open (NO), and normally closed (NC). You can connect the load between the COMM and NO or between the COMM and NC terminals. When you connect between COMM and NO, the output remains disconnected when the relay is turned off and connected when the relay is turned on.

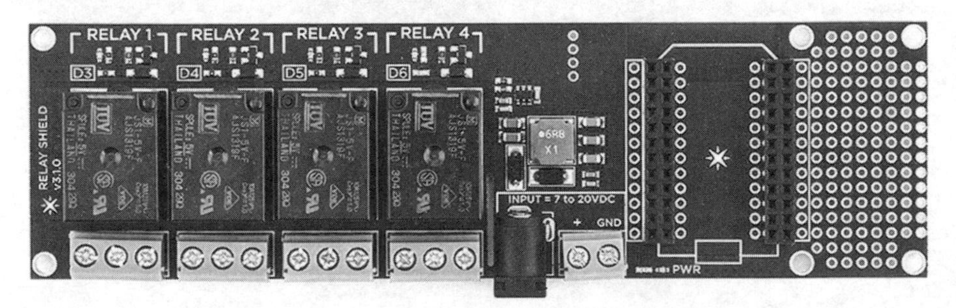

Figure 7.3 *Relay shield.*

Description	Appendix
Photon board	M1
Relay shield	M3
Power supply	H7
Lightbulb	H8
9-V battery	H9
Equipment wire	H10

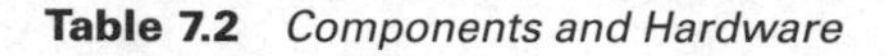

Table 7.2 *Components and Hardware*

The relay shield uses a switch mode regulator that provides a constant 5 V to the Particle and the relays. The regulator is rated at 1.2 A maximum output current, which is sufficient to power the Photon and the four relays while still having power left over to control other things that you connect to the Photon device. You can power the relay shield using the DC socket or through a screw terminal with a voltage range between 7 and 20 V DC.

Let's take a look at some code that we can use to control a simple lightbulb. The relay acts as our switch, which is normally open, and when pin D3 on the Photon is turned HIGH, it activates the relay and switches on the light. The hardware for this experiment can be seen in Table 7.2 and the layout diagram in Figure 7.4.

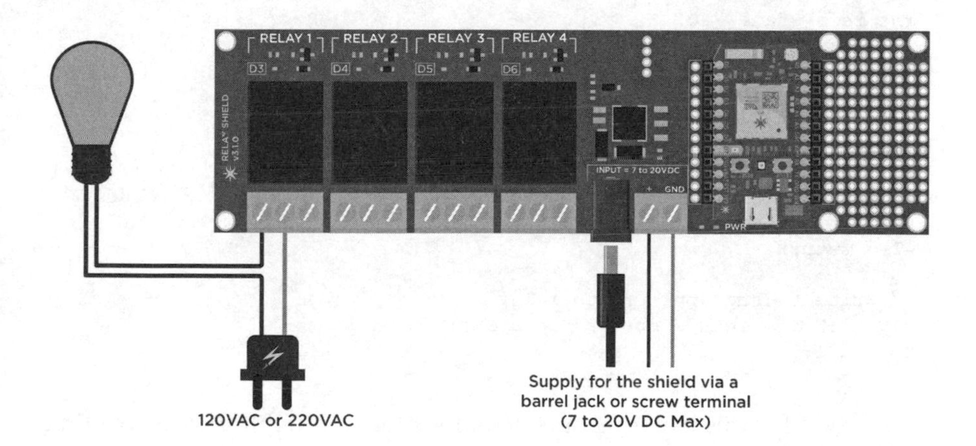

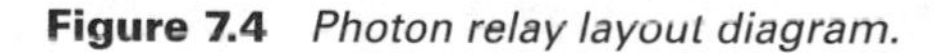

Figure 7.4 *Photon relay layout diagram.*

Here is the code to switch on the lightbulb:

```
int RELAY1 = D3;
int RELAY2 = D4;
int RELAY3 = D5;
int RELAY4 = D6;

void setup()
{
   //Initialize the relay control pins as output
   pinMode(RELAY1, OUTPUT);
   pinMode(RELAY2, OUTPUT);
   pinMode(RELAY3, OUTPUT);
   pinMode(RELAY4, OUTPUT);
   // Initialize all relays to an OFF state
   digitalWrite(RELAY1, LOW);
   digitalWrite(RELAY2, LOW);
   digitalWrite(RELAY3, LOW);
   digitalWrite(RELAY4, LOW);

   //register the Spark function
   Spark.function("relay", relayControl);
}

void loop()
{
   // This loops forever
}

// command format r1,HIGH
int relayControl(String command)
{
  int relayState = 0;
  // parse the relay number
  int relayNumber = command.charAt(1) - '0';
  // do a sanity check
  if (relayNumber < 1 || relayNumber > 4) return -1;

  // find out the state of the relay
  if (command.substring(3,7) == "HIGH") relayState = 1;
  else if (command.substring(3,6) == "LOW") relayState = 0;
  else return -1;

  // write to the appropriate relay
  digitalWrite(relayNumber+2, relayState);
  return 1;
}
```

The code for this experiment is actually pretty straightforward. We first set up each of the relays and assign the Photon pin number to the relay:

```
int RELAY1 = D3;
int RELAY2 = D4;
int RELAY3 = D5;
int RELAY4 = D6;
```

We then set the digital pins on the Photon to outputs, as when we switch the pin to HIGH or LOW; this will also switch the relay ON or OFF:

```
pinMode(RELAY1, OUTPUT);
pinMode(RELAY2, OUTPUT);
pinMode(RELAY3, OUTPUT);
pinMode(RELAY4, OUTPUT);
```

When we first run the program, it is best practice to make sure all the relays are switched off before we do anything:

```
digitalWrite(RELAY1, LOW);
digitalWrite(RELAY2, LOW);
digitalWrite(RELAY3, LOW);
digitalWrite(RELAY4, LOW);
```

If we want to control all the relays over the Internet using a Web browser, then we need to register a Spark function:

```
Spark.function("relay", relayControl);
```

In the actual function we do a number of checks, such as determine which relay button was pressed and whether or not the relay is HIGH or LOW to determine the next state of the relay. Once the program has done that, we can write to the digital pin:

```
int relayControl(String command)
{
  int relayState = 0;
  // parse the relay number
  int relayNumber = command.charAt(1) - '0';
  // do a sanity check
  if (relayNumber < 1 || relayNumber > 4) return -1;

  // find out the state of the relay
  if (command.substring(3,7) == "HIGH") relayState = 1;
  else if (command.substring(3,6) == "LOW") relayState = 0;
  else return -1;

  digitalWrite(relayNumber+2, relayState);
```

The Hypertext Markup Language (HTML) code for controlling the relays is as follows:

```
<html>
<head>
<script src="http://ajax.googleapis.com/ajax/libs/jquery/1.3.2/jquery.
min.js" type="text/javascript" charset="utf-8"></script>

<script>
var accessToken = "cb8b348000e9d0ea9e354990bbd39ccbfb57b30e";
var deviceID = "54ff72066672524860351167"
```

```
var url = "https://api.spark.io/v1/devices/" + deviceID + "/relay";
function setRelay(message)
{
       $.post(url, {params: message, access_token: accessToken });
}
</script>
</head>

<body>

<h1>Relay Control</h1>
<table>
<tr>
       <td><input type="button" onClick="setRelay('11')" value="Relay
1 ON"/></td>
       <td><input type="button" onClick="setRelay('10')" value="Relay
1 OFF"/></td>
</tr>
<tr>
       <td><input type="button" onClick="setRelay('21')" value="Relay
2 ON"/></td>
       <td><input type="button" onClick="setRelay('20')" value="Relay
2 OFF"/></td>
</tr>
<tr>
       <td><input type="button" onClick="setRelay('31')" value="Relay
3 ON"/></td>
       <td><input type="button" onClick="setRelay('30')" value="Relay
3 OFF"/></td>
</tr>
<tr>
       <td><input type="button" onClick="setRelay('41')" value="Relay
4 ON"/></td>
       <td><input type="button" onClick="setRelay('40')" value="Relay
4 OFF"/></td>
</tr>
</table>
</body>
</html>
```

Programmer Shield

The programmer shield is for the very advanced users who want to gain full access and control over the Photon board. This shield is an FT2232H-based JTAG programmer shield that is fully compatible with OpenOCD and Broadcom's WICED integrated development environment (IDE). The FT2232H chip is set up to provide a USB JTAG and USB universal asynchronous receiver/transmitter (UART) interface for the Photon board. It can also be

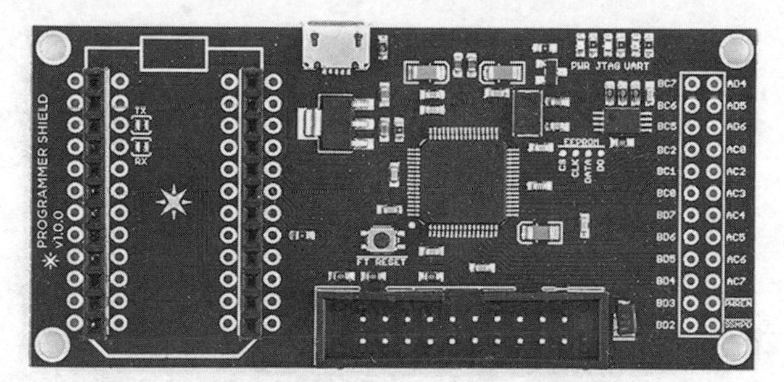

Figure 7.5 *JTAG programming shield.*

configured by the user by reprogramming the on-board config electrically erasable programmable read-only memory (EEPROM). The unused pins are marked and broken out on to easy-to-access headers as shown in Figure 7.5.

Power Shield

The power shield does exactly that—it allows you to power your Photon from other power sources than USB. The shield has an intelligent battery charger and power management unit, along with a wide-range input voltage regulator and an (inter-IC) I2C-based power gauge. You can power a Photon with either a USB plug or a DC power supply from anywhere between 7 and 20 V DC and can also charge a 3.7-V lithium ion–polymer (LiPo) battery at the same time. This is the perfect companion to your Photon board, giving you the ability to truly create mobile applications from anywhere in the world. The power shield is shown in Figure 7.6.

The power shield system switches between the different power sources automatically, reducing the charge and discharge cycle stress on the battery. The fuel gauge allows you to easily monitor the battery's state of charge, allowing it to notify the user remotely over the Internet and take the necessary actions when needed. The shield is also set up so that when powered from a USB port as well as a DC power source, the shield will automatically choose the DC source over USB. The charge current is set to 500 mA when charging from USB and it is set to 1 A when charging from a DC power source.

Figure 7.6 *Particle power shield.*

The Internet Button

The Internet button is a great way to get started with the Internet of Things by creating interactive projects using a number of inputs and outputs that feature on the shield. Not only that, but it is a simple way to start building your very own prototypes straight out of the box. The Internet button lets you play with lots of light-emitting diodes (LEDs) and buttons, as well as an accelerometer. The Internet button, shown in Figure 7.7, has 11 individually controllable RGB (red, green, blue) LEDs as well as four tactile buttons

Figure 7.7 *Particle Internet button.*

underneath the board. The Internet button can be powered through USB or externally through a DC power source from 3.6 to 6 V DC.

This item includes the following features:

- Push buttons to control hundreds of Web services
- On-board LEDs to display data or alerts
- Completely assembled—no wiring or soldering
- No coding required
- Pluggable headers for adding sensors
- Helper functions makes complex behavior simple
- Open computer-assisted design (CAD) model for 3D printing of custom covers
- Rest application programming interface (API) for rapid development

The Internet button comes with some great examples to get you started using the features of the Internet button, such as the LED and push buttons. If you have purchased an Internet button, it comes supplied with a Photon board, so no additional hardware is required to use these examples.

Grove Starter Kit for Photon

The Grove Starter Kit for Photon is an easy-to-use, plug-and-play kit for the Photon, as shown in Figure 7.8.

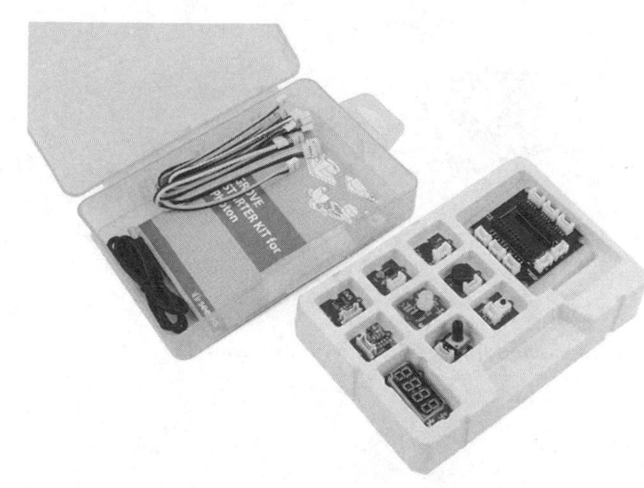

Figure 7.8 *Grove Photon kit.*

The base shield lets you connect a variety of included Grove modules. It includes a Grove Shield for Photon and several other Groves. Grove Shield for Photon is an expansion board, which is easier for users to build Grove's four-pin standard interfaces from the Spark Photon. It also helps the user avoid using too many cables to build the prototype. The kit includes the following modules:

- Grove Shield for Photon
- Grove—Button
- Grove—Buzzer
- Grove—Rotary Angle Sensor
- Grove—Temperature Sensor
- Grove—Light Sensor
- Grove—Chainable RGB LED
- Grove—Three-Axis Digital Accelerometer (±1.5 g)
- Grove—Four-Digit Display
- Grove—Vibration Motor
- User's Manual

The Grove modules can easily be connected to the Photon board, as shown in Figure 7.9.

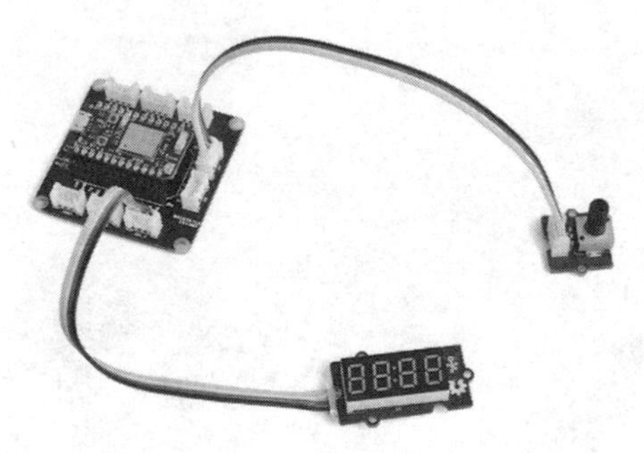

Figure 7.9 *Grove modules connected to the Photon shield.*

Adafruit Particle NeoPixel Ring Kit

Add some dazzle to your Photon with this custom-made NeoPixel ring kit! Twenty-four ultra-bright smart LED NeoPixels are arranged in a circle with 2.6-inch outer diameter, as shown in Figure 7.10.

Snap in your Photon and upload the NeoPixel library code to light up the LEDs—make an Internet of Blinky! Each LED is addressable, as the driver chip is inside the LED. Each one has ~18-mA constant current drive so the color will be consistent even if the voltage varies, and no external choke resistors are required, making the design slim. Power the whole thing with a 3.5- to 5.5-V DC battery pack and you're ready to rock.

To make your project portable, there is a Japan Solderless Terminal (JST) connector for attaching an external battery. Power with 3.5- to 5.5-V DC—a rechargeable LiPo battery works great, or a 3 × AAA or 3 × AA battery pack. The JST included enables you to make your own battery connection. Use pin D6 for the NeoPixel library code—all other pins are available for other uses and have two breakouts on either side so you can wire up other sensors or devices.

The ring kit comes as a single, round printed circuit board (PCB) with 24 individually addressable RGB LEDs assembled and tested, two 12-pin

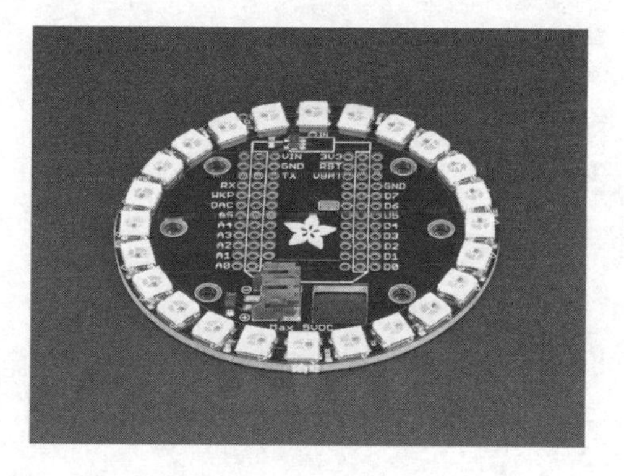

Figure 7.10 *Adafruit NeoPixel Photon ring.*

0.1-inch socket headers, and a bonus JST cable. Some light soldering is required—you can solder the two sockets in place to allow unplugging of the Photon, or just solder it directly in place for a slimmer look.

Summary

Hopefully you can see that there are many ways in which your projects and experiments can be simplified by using some of these add-on boards and shields. In the next chapter we will take a look at using the "if this then that" (IFTTT) Web service with your Photon board.

8

IFTTT

IFTTT stands for "if this then that" and does exactly that. IFTTT is a Web service that allows you to link real-time information to your Photon board. The platform works by users creating recipes such as "if my favorite football team scores, then send me an e-mail." IFTTT is also well integrated to popular Web platforms such as Twitter, Facebook, Google Mail, and many more. Moreover, the IFTTT platform can integrate with our Photon board such that if some action is triggered, such as temperature, motion, or any data from other sensors, then you can define what to do next, such as sending a tweet or an e-mail to someone. This is the basic principle of how the IFTTT platform works, and next we'll be taking a closer look at how we can connect this service to our Photon board.

If This Then That

Before we get started, you need to register an account with IFTTT, which is completely free, from https://ifttt.com/. Once you have registered your details, you can get started with writing your very own recipes to do all sorts of weird and wonderful things. IFTTT also give users options to share recipes so you, too, can share your recipes with the world—after all, we should be great believers in open source.

As previously mentioned, IFTTT integrates into a lot of Web services such as Facebook, Twitter, and Google Mail. If you want to use these services in your recipes, you need to grant IFTTT access to these through the APIs. Feel free at this point to explore all the other applications that are available to you, as well as read some getting started instructions, which also would be useful—maybe try out a few of the existing recipes before we get started to get a feel for how it works.

Sunrise E-mail Alert

This project uses the same fundamentals as in Chapter 5. In that chapter we looked at using a photocell as an input device using the `analogRead()` function. If you haven't done so already, I would highly recommend referring back to this chapter before proceeding. If you have purchased the Photon kit, then you should have received some resistors and a photoresistor, among other components. You can actually wire this circuit up without any jumper wires if you do not have any at hand by using one of the analog pins as a 3.3-V power pin and turning the output to HIGH, giving an output of 3.3 V; this also gives us a stable voltage reading because it does not fluctuate like the 3.3-V pin does sometimes.

The hardware for this experiment is shown in Table 8.1 and the breadboard layout diagram in Figure 8.1.

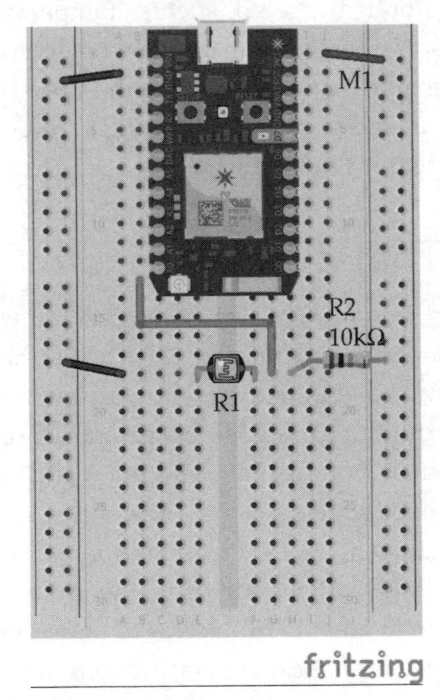

Figure 8.1 *Breadboard layout diagram for the photocell.*

Schematic Reference	Description	Appendix
M1	Photon board	M1
	Breadboard	H1
	Jumper wires	H2
R1	Photocell	R4
R2	10-K resistor	R3

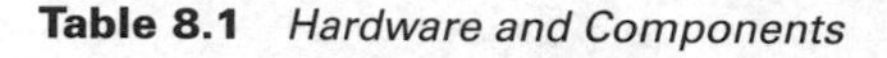

Table 8.1 *Hardware and Components*

Now that we have our circuit built, we need to write our program to the Photon board so we can read our variable using IFTTT. Here is the code for our experiment:

```
int photoresistor = A0;

int power = A5;

int analogvalue;


void setup() {


    pinMode(photoresistor,INPUT);
    pinMode(power,OUTPUT);

    digitalWrite(power,HIGH);

    Spark.variable("analogvalue", &analogvalue, INT);

}

void loop() {
    analogvalue = analogRead(photoresistor);
}
```

The code for our experiment is really simple. We set pin A5 as an output, so we apply a voltage through the circuit, then we read the value from the photoresistor and store the value in a variable called `analogvalue`. Remember that the values we are reading from the analog pin range from 0 to 4095.

Now that we have everything set up on our Photon board, we can go ahead and create a new recipe on our IFTTT account by clicking the Create A New Recipe icon as shown in Figure 8.2.

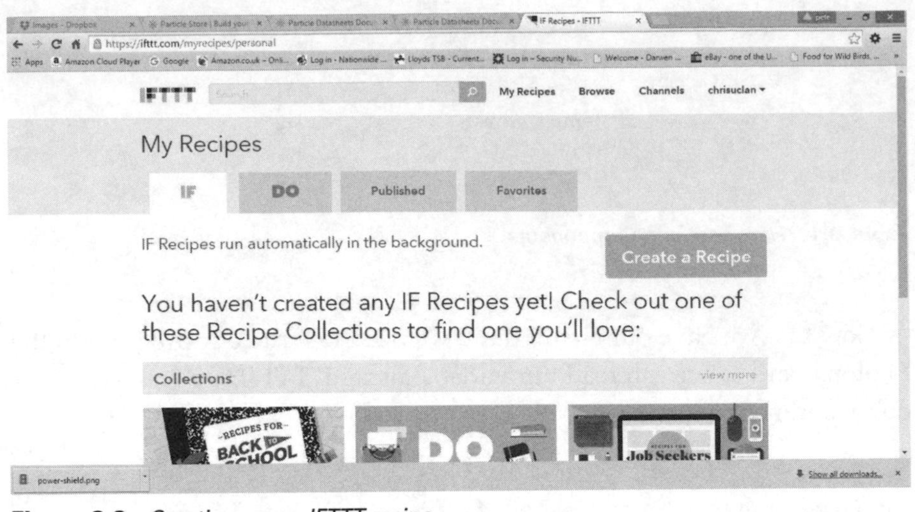

Figure 8.2 *Creating a new IFTTT recipe.*

You will be presented with a screen that shows the words "if this then that" with "this" underlined as a hyperlink. This gives us an indication that we need to define what "this" is by clicking the hyperlink. We can then set a trigger, which will determine how we run our recipe. We know that when a light is shone on the photocell we want to trigger an e-mail to say something. The next screen takes you to a long list of icons from a wide variety of Web services, which we can select to act as a trigger. Scroll down the list until you see the Particle logo, or alternatively you can use the search box and type "particle," which will show the Particle app as seen in Figure 8.3.

Click the Particle app, and you will be prompted to enter your Particle cloud login details if this is your first time using IFTTT. This will grant you access to use the particle.io features, so enter both your username and password here. Once granted access, you will be presented with a trigger option on the next page, as shown in Figure 8.4. These triggers determine how we interact with our Particle and how we process the result the Particle returns to us. We know from writing our program that the levels of light using the photocell are stored in a variable called `analogvalue` so we know that we can choose Monitor A Variable.

Create Recipe - IFTTT
https://ifttt.com/myrecipes/personal/new
Nike+
NPR
Numerous
Office 365 Calendar
Office 365 Contacts
Office 365 Mail
openHAB
Parrot Flower Power
Particle
People
Phone Call
Pinboard
Pinterest
Pocket
Pryv
Qapital
Qblinks Qmote
QualityTime

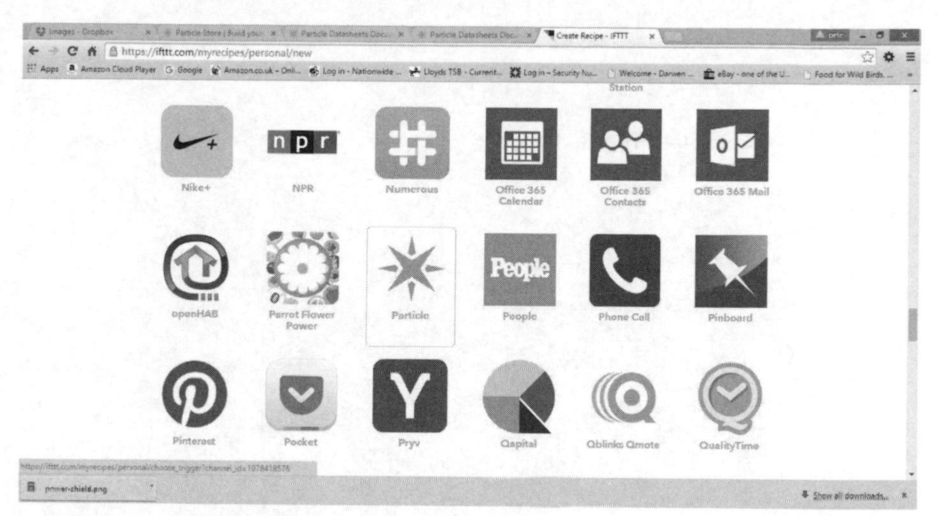

Figure 8.3 *Particle app in IFTTT.*

Make sure your Photon is connected and the program that we uploaded previously is running. When you click the drop-down box under the IF (variable name) field, you should see a list of all available variables you have created on the Photon board, as shown in Figure 8.5.

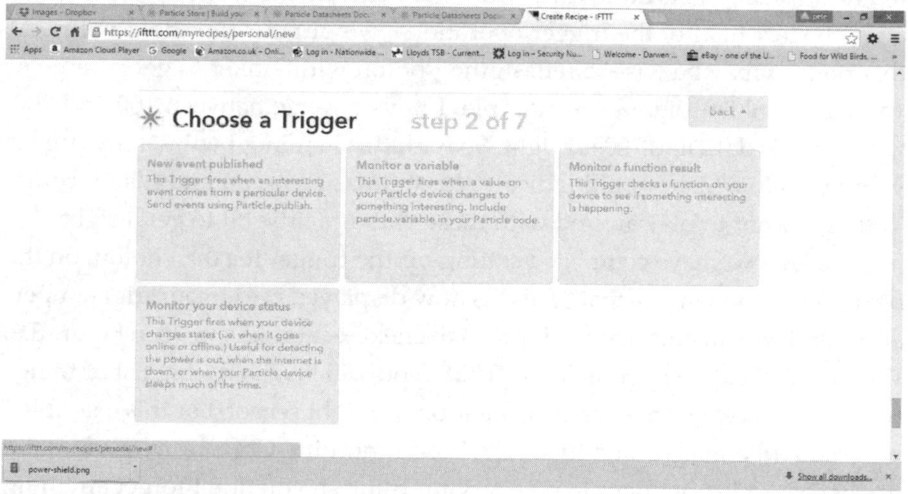

Create Recipe - IFTTT
https://ifttt.com/myrecipes/personal/new
Choose a Trigger
step 2 of 7
New event published
This Trigger fires when an interesting event comes from a particular device. Send events using Particle.publish.
Monitor a variable
This Trigger fires when a value on your Particle device changes to something interesting. Include particle.variable in your Particle code.
Monitor a function result
This Trigger checks a function on your device to see if something interesting is happening.
Monitor your device status
This Trigger fires when your device changes states (i.e. when it goes online or offline.) Useful for detecting the power is out, when the internet is down, or when your Particle device sleeps much of the time.

Figure 8.4 *Particle triggers.*

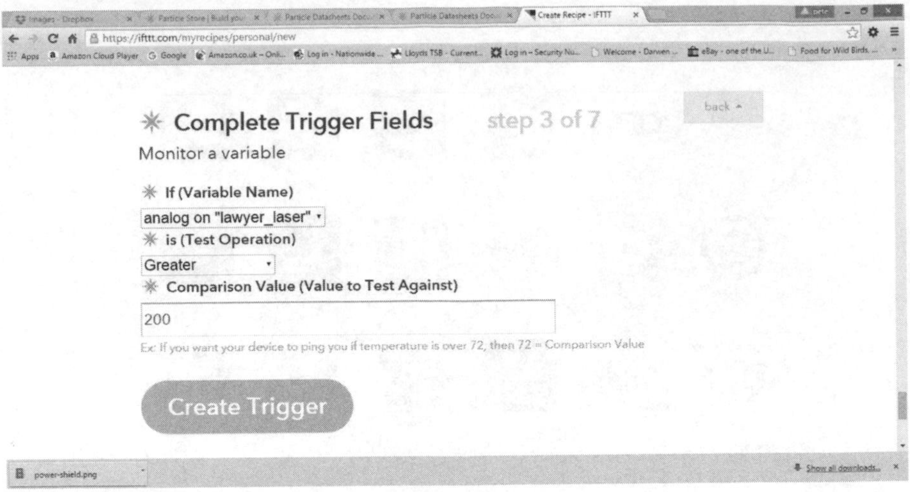

Figure 8.5 *Configuring Photon triggers.*

In the next field you will see the test operation; this is where you can apply some simple math to determine how you want the trigger to work. In our case we want to trigger the e-mail when the value of light is above a certain value, indicating there is some light; therefore, we want the test operation to be greater than the value we set in the next step. Alternatively, if you wanted to know when it gets dark, then you would change the value to less than.

In the last field of the trigger operations, we put our threshold value. At this point it may be wise to reflash the Photon with Tinker to get a range of values of ambient light—for example, I have a range between 100 and 700, where 100 is dark and 700 is bright. So with this in mind, I will set my trigger value to 200, which should indicate the sun has come up and it's morning. You may want to play around with these values a little bit to get it right.

Now that we have completed setting up the trigger for the Photon, on the next screen you will see that "This" is now displayed next to our trigger operation, and we can now define "That" in the next section, as shown in Figure 8.6. When you click the hyperlink for "That" you will be given a big list of things we can do once we have hit the trigger on our light sensor. For this example I will send an e-mail to myself to say "Good morning Chris" —you can quite easily write a list of things to do every morning so you don't forget anything before you leave for work or when you get to work. If you have a Google Mail

Figure 8.6 *Define an action.*

account, you can select the Gmail icon in the list, as shown in Figure 8.7; alternatively, if you do not have a Gmail account, you can use IFTTT's internal mail server, which will use the e-mail address that you provided when you signed up for the IFTTT account.

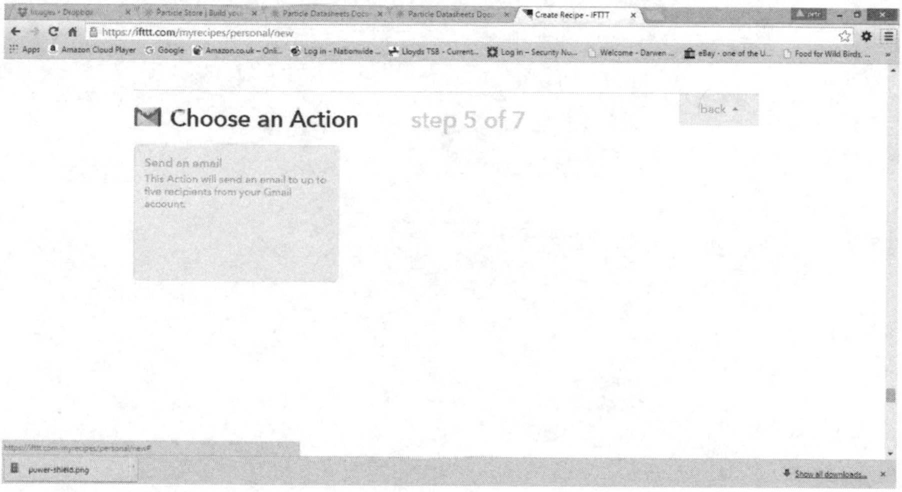

Figure 8.7 *List of actions.*

If you have selected Gmail to send an e-mail, then you will be prompted to log in to your Gmail account to grant access to its features—you only have to do this once, as the log-in details will be saved for future recipes. The next step shows you a list of actions that can be completed under that particular Web service. Gmail only allows you to send an e-mail, and as such you can click this to proceed to the next step. In the next step you can complete all the action fields, such as the e-mail address to send the e-mail to; you can select up to five e-mail addresses, separating them by commas. You can also change the subject of the e-mail as well as the content of the e-mail, and if you really want to, you can upload a file to send as well, as shown in Figure 8.8.

In the body of the e-mail you can include certain variables that are not filtered as plain text; they are as follows:

- **CreatedAt** The date and time that the e-mail was created and sent
- **DeviceName** The device name of the Photon board that you provided
- **Value** The actual value of the variable when it was triggered
- **Variable** The name of the variable that we are reading; in this case it is `analogValue`

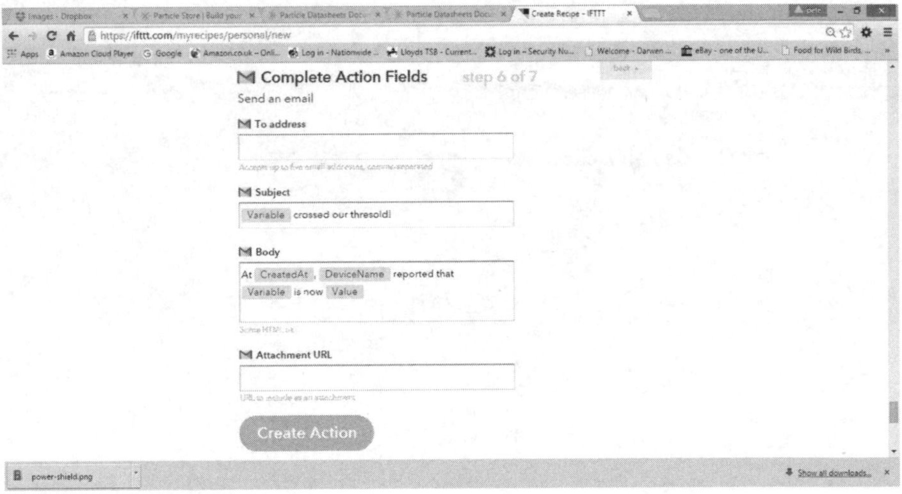

Figure 8.8 *Gmail actions.*

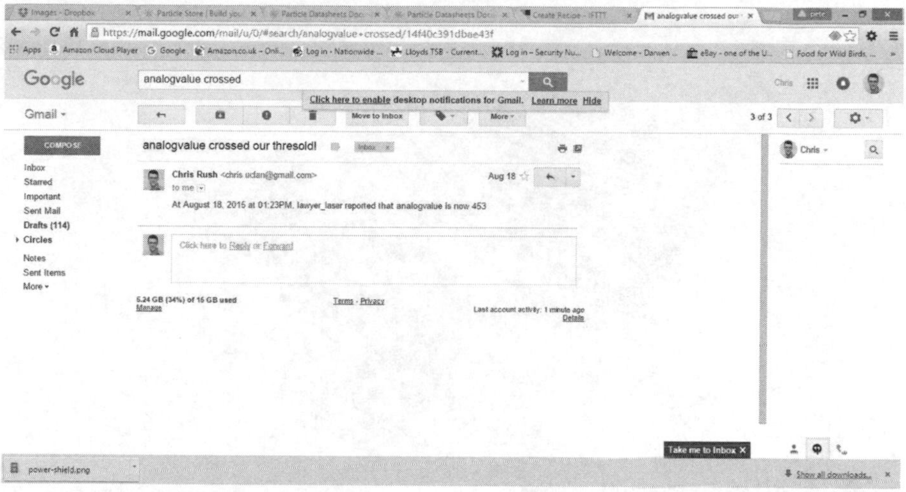

Figure 8.9 *IFTTT trigger e-mail.*

Once you have filled in all the required fields, click Create Action to go to the final step of the recipe. Here you will click Create And Activate, which will create the recipe and have it go live on the IFTTT Web service. The live recipe will then check the Photon every 15 minutes—this makes sure you do not get flooded with e-mails every second and balances out the server load.

Try placing the sensor in a bright environment to test the IFTTT recipe you created. After about 15 minutes you should receive an e-mail like that shown in Figure 8.9 with all the details that you filled out in the e-mail actions section.

You can, of course, trigger all sorts of actions from the Photon board, and you can easily go back and change the settings of a recipe instead of creating a new one.

Create a Twitter Alert Using Grove Modules

Now that we have looked at using our Photon to trigger an event using IFTTT, we are going to look at how we can use a Web service such as Twitter to trigger something to happen using the Photon board. For this experiment we are going to be using the SeeedStudio starter kit for the Photon. Grove is a modular, ready-to-use set of tool blocks, as shown in Figure 8.10.

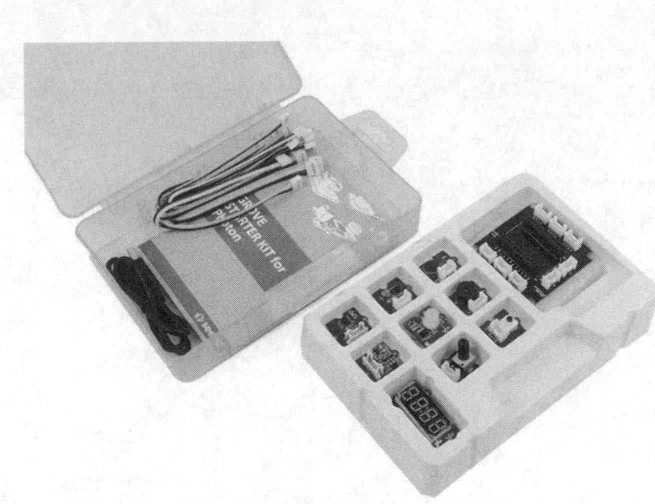

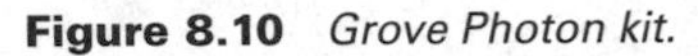

Figure 8.10 *Grove Photon kit.*

Its uses a building block approach to electronics, which is great for users with little to no experience of electronic circuits. The Grove system consists of a base shield and some modular blocks; in our instance, the Photon kit comes supplied with the following parts:

- Grove Shield for Photon
- Grove—Button
- Grove—Buzzer
- Grove—Rotary Angle Sensor
- Grove—Temperature Sensor
- Grove—Light Sensor
- Grove—Chainable RGB (red, green, blue) light-emitting diode (LED)
- Grove—Three-Axis Digital Accelerometer (±1.5 g)
- Grove—Four-Digit Display
- Grove—Vibration Motor
- User's Manual

Figure 8.11 *Photon base shield.*

Description	Appendix
Photon board	M1
Grove Particle shield	M4
Grove vibration motor	M4
Grove buzzer	M4

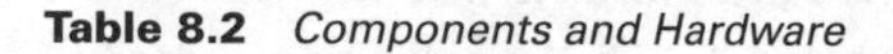

Table 8.2 *Components and Hardware*

In this experiment using IFTTT, every time someone mentions you on Twitter, the buzzer and vibration motor will go off for one second, alerting you to the tweet. To get started, go ahead and insert the Photon board into the Grove Photon shield as shown in Figure 8.11, making sure the orientation is correct.

For this experiment we are going to be using the Grove buzzer and the Grove vibration motor. Connect the Grove vibration motor to connector A4 on the Particle shield, and connect the buzzer to the I2C_2 port on the shield. The hardware for this experiment can be found in Table 8.2.

The software for this experiment is simple. We create a function that IFTTT passes through a value. We then check the value to see if it matches and then indicate what sound we want (such as a buzzer) and turn the motor on. The software for this experiment is as follows:

```
#define MOTORPIN A4
#define BUZZPIN D1


void setup() {
    pinMode(MOTORPIN, OUTPUT);
    pinMode(BUZZPIN, OUTPUT);
    Spark.function("Twitter", twitter);
}
```

```
void loop() {

}

int twitter(String command)
{
    if (command == "buzz")
    {
        digitalWrite(MOTORPIN, HIGH);
        digitalWrite(BUZZPIN, HIGH);
        delay(1000);
        digitalWrite(MOTORPIN, LOW);
        digitalWrite(BUZZPIN, LOW);
        return 1;
    }
    else return -1;

}
```

First we need to define the pins that we are going to use on the Photon board. The buzzer is connected to digital pin 1, and the vibration motor is connected to analog pin 4.

```
#define MOTORPIN A4
#define BUZZPIN D1
```

Next in the `setup` function we set both the buzzer and the motor to outputs and create a Spark function that we can pass data though from IFTTT. Remember that the Spark function name cannot be longer than 16 characters.

```
pinMode(MOTORPIN, OUTPUT);
pinMode(BUZZPIN, OUTPUT);
Spark.function("Twitter", twitter);
```

We ignore the `loop` function because when the Spark function is called, it initializes the `twitter` function. The Spark function passes through a string value called `command`, and we check to see if this value is equal to "buzz," then write the motor and buzzer to HIGH for one second before turning them off. If the function does not equal the value of "buzz," then the function returns –1 as failed.

```
int twitter(String command)
{
    if (command == "buzz")
    {
```

```
        digitalWrite(MOTORPIN, HIGH);
        digitalWrite(BUZZPIN, HIGH);
        delay(1000);
        digitalWrite(MOTORPIN, LOW);
        digitalWrite(BUZZPIN, LOW);
        return 1;
    }
    else return -1;
}
```

Now that we have our Photon set up and running, we can now turn our efforts to the IFTTT Web service. Log in to your account and create a new recipe. This time we are going to use Twitter as a trigger rather than the Photon as before. At this point you may be prompted to authenticate your Twitter account and asked to give permission for IFTTT to use the Twitter Web service. On the trigger page there are quite a number of different types of triggers that you can choose from—the one we will use for this experiment is "Twitter mentions of you." Whenever your Twitter ID is tagged in a tweet, this will trigger the events in the recipe. Select New Mention Of You and proceed to the next step. Here there is nothing to add, so click Create Trigger. Now we can move on to the next step of setting up the "That" section of the recipe.

Click the Particle icon from the list of action sources, and you will see a list of actions that can be done with your Photon, as seen in Figure 8.12.

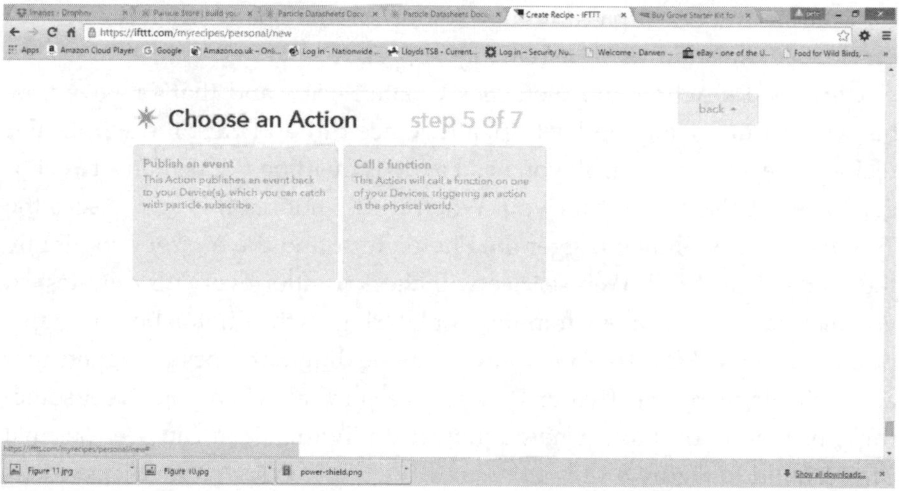

Choose an Action step 5 of 7

back

Publish an event
This Action publishes an event back to your Device(s), which you can catch with particle subscribe.

Call a function
This Action will call a function on one of your Devices, triggering an action in the physical world.

Figure 8.12 *Photon actions.*

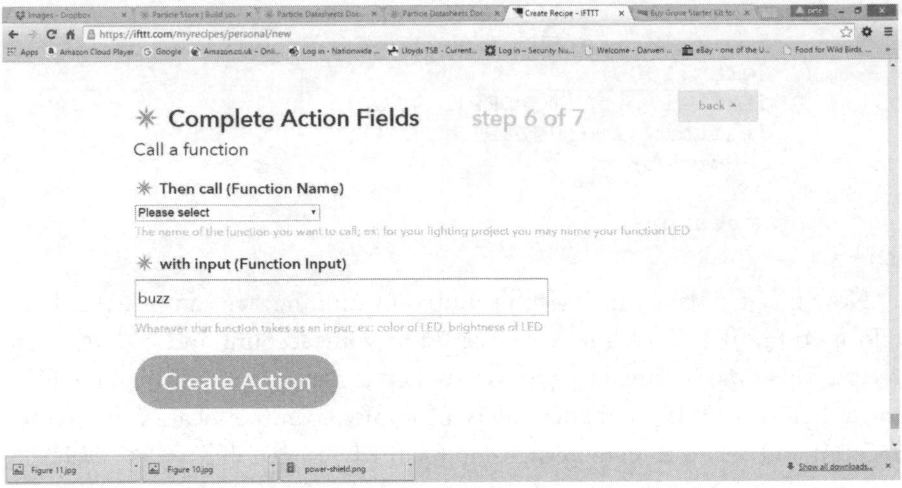

Figure 8.13 *Photon action fields.*

Select the option Call A Function, and you will be asked to complete a list of action fields to set up the action. From the Then Call drop-down list you should see a list of functions that have been set up on your Photon board, if it is connected. You should see in this drop-down box a function called Twitter, which we set up in our code on the Photon. Delete the contents of the action field "with input" and type in this field "buzz," which is going to be the value that we will pass through to our Photon board when the Twitter service has triggered, as shown in Figure 8.13.

Click Create Action and then click Create Recipe and that's it—we now have everything ready and set up; all we need to do is test the scripts and recipe. To test the project, all you need to do is mention yourself in a tweet or get someone else to do it for you. When IFTTT notices this and checks the Twitter feed, it will then trigger the Photon to sound the buzzer and vibrate the motor. The IFTTT Web service will check Twitter every 15 minutes, so you may have to wait a few minutes until it triggers the Photon board. At this point you can feel free to play around with the different types of triggers that are available using the Twitter Web service, such as when you tweet something or when you have a new follower on Twitter. You can see the final project build in Figure 8.14.

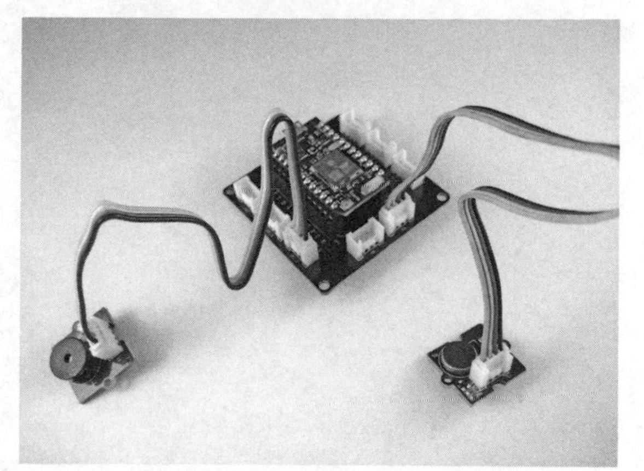

Figure 8.14 *Grove Photon connected to buzzer and vibration modules.*

Summary

In this chapter we have looked at using that the IFTTT Web service to control electronics on the Photon board, as well as using the Photon board to control some Web services such as Twitter. There are many other services that you can use, such as Facebook, e-mail, or even some sports results, that can trigger LEDs or something similar on the Photon board.

9

Troubleshooting Your Device

Sometimes things don't always go according to plan, and it can be a hard task to find out what went wrong and how to fix it. Luckily, the Photon board has an RGB (red, green, blue) light-emitting diode (LED) placed in a central position on the board that can indicate a number of things based on the color of the LED and the number of times it blinks, or doesn't, as the case may be.

Device Modes

These modes are the typical behaviors of the Photon board that you will see on a regular basis; they are the light patterns that determine exactly what your Photon board is doing, which is great when trying to work out what has happened when something goes wrong.

- **Connected** When the Photon board is pulsing cyan, your device is connected to the Internet. When the Photon is in this mode, you can call functions and flash code to the board.
- **Firmware update** If your device is flashing magenta, it is currently loading an app or updating its firmware. This device mode is triggered by a firmware update or by flashing code you have written from the Particle Dev or Particle Build. You will also see this mode when you connect your Photon to the cloud for the first time.

NOTE If you enter this mode when holding down the SETUP button on boot, the flashing magenta indicates that letting go of the SETUP button will enter Safe mode to connect to the cloud and not run application firmware.

- **Looking for the Internet** If the Photon board is flashing green, it is trying to connect to the Internet. If you have already entered your Wi-Fi settings, then the Photon board will flash for a few seconds before connecting and then start pulsing cyan. If you have not yet connected your device to the Wi-Fi network, you need to set your device to Listening mode.
- **Connecting to the cloud** When the Photon board is in the process of connecting to the cloud, it will rapidly flash cyan. You will most likely see this happen when you first connect your Photon board to the network after it has flashed green.
- **Wi-Fi off** If the Photon board LED is pulsing white, then the Wi-Fi module is off. There are a couple of reasons why this may occur:
  - You have set your module to `manual` or `semi_automatic` in your user firmware
  - You have called `wifi.off()` in your user firmware program
- **Listening mode** This is probably the most important mode you will encounter. When the Photon board is in Listening mode, it is waiting for your input to connect to the Wi-Fi network. The Photon needs to be in Listening mode in order to begin connecting with the smart phone app or over universal serial bus (USB). To put the Photon board in Listening mode, simply hold down the SETUP button for three seconds until the RGB LED begins to flash blue.
- **Wi-Fi network reset** To erase the Wi-Fi network on the Photon board, hold down the SETUP button for about 10 seconds until the RGB LED rapidly flashes blue. Alternatively, you can hold down the SETUP button and tap the RESET button while doing so until the RGB LED turns white.
- **Safe mode** Safe mode connects the device to the cloud but does not run any application firmware program. This mode is one of the most useful for development or for troubleshooting. If something has gone wrong with the app or firmware you have loaded onto your Photon

board, you can set your device to Safe mode. This runs the device's system firmware but doesn't execute any application code, which can be useful if the application code contains bugs that stop the device from connecting to the cloud. The Photon board indicates that it is in Safe mode when the LED blinks magenta. To put your Photon board in Safe mode, hold down both buttons in the board and release the RESET button while keeping the SETUP button held down until it flashes magenta release the setup button. The Photon board will then enter Safe mode if there is no application code flashed to the board or when the application firmware is invalid.

- **Device firmware upgrade** If you want to program your Photon board with a custom firmware through USB, you will need to use this mode. This mode triggers the on-board bootloader that accepts firmware binary files through the dfu utility program. To enter the Device Firmware Upgrade mode, you need to hold down both buttons and release the RESET button while keeping the SETUP button held down. Wait until the LED starts flashing yellow and then release the SETUP button.

Troubleshooting Modes

- **Wi-Fi module not connected** If the Wi-Fi module is on but not connected to the network, your Photon with blink blue. This will be a dark blue, not a cyan color.
- **Cloud not connected** When the device is connected to a Wi-Fi network but not to the cloud, it will blink green.
- **Bad public key** When the server public key is bad, the Photon board will flash alternating colors of cyan and red. A red flashing LED can indicate the following errors:
  - **Two red flashes** Could not reach the Internet
  - **Three red flashes** Connected to the Internet but could not reach the Particle cloud
  - **Flashing orange** Bad device keys
- **Red flash** A pattern of more than 10 red flashes indicates that the firmware has crashed. The pattern flashes in an SOS sequence with three short flashes, three long flashes, and three short flashes again.

If you encounter this issue, refer back to the Safe mode option and try to reflash the firmware again. A number of other codes may be expressed after the SOS red flashes:

- Hard fault
- Nonmaskable interrupt fault
- Memory manager fault
- Bus fault
- Usage fault
- Invalid length
- Exit
- Out of heap memory
- Serial Peripheral Interface (SPI) overrun
- Assertion failure
- Invalid case
- Pure virtual call

Summary

This brief chapter should give you a good understanding of what is happening to your Photon in terms of the RGB LED flashes and colors. You should be able to adjust things when everything does not quite go according to plan and understand what each mode is doing and why. You can also visit the Particle community pages for more help if you are still struggling with getting something to work. The community is supportive and active in answering all kinds of technical questions.

Tools and Tips

This section gives you some useful tips for creating your own projects and how to best utilize your resources. Starting your own projects can be a bit daunting at first and sometimes can seem frustrating and complicated—this useful information should help you on your way.

Breadboards and Prototyping Boards

A breadboard is usually a rectangular plastic acrylonitrile butadiene styrene (ABS) box with lots of little holes in it; the holes are contacts in which you can insert electrical components or wires into with ease. Breadboards are often used to put together a concept design of a circuit without having to solder any components. Instead you just poke the wires or legs of the component into the holes, creating a contact. The contacts are usually arranged in rows by connecting the metal contacts underneath the breadboard. The best thing about using a breadboard is that you can change the circuit design at any point so you can replace or rearrange components with ease without having to solder/desolder any joints.

When you place components in a breadboard, not much happens unless you connect jumper wires to create an electrical circuit. Wire used in electronics is copper surround by an outer plastic insulation, usually called hook-up wire. Wire comes in all sorts of diameters, often referred to as gauge; the standard measurement in the United States is American wire gauge (AWG). It is always advisable to used solid wire rather than stranded wire

Figure A.1 *Breadboard.*

because solid wire inserts into the breadboard much easier than does stranded wire. If you are lucky, your electronics shop will sell jumper wires, which are short lengths of wire with a single pin on each end.

If you create your circuit design on a breadboard, you may decide that you want to make it permanent by soldering components in place on a printed circuit board (PCB). To do this, you may have to get a universal printed circuit board, which in some ways is similar to a breadboard layout. A prototyping printed circuit board has rows of individual holes across them, much like all the pins on a breadboard. Generally, all the components will go on the top

Figure A.2 *Jumper wires.*

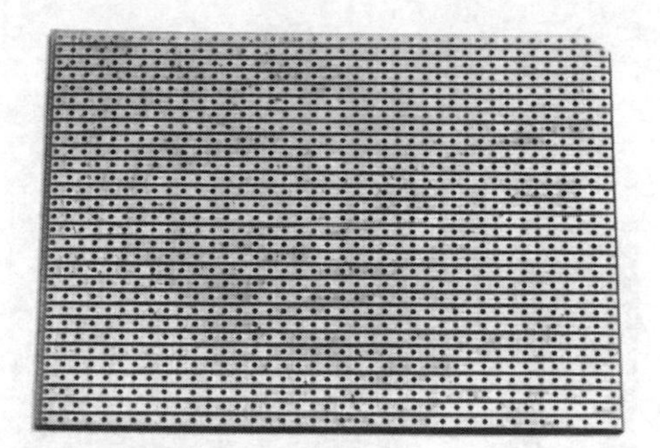

Figure A.3 *Prototyping board.*

of the board, and you solder underneath, and when you solder all the wires, they usually go underneath also. This makes a much neater and cleaner board to work with and can help avoid a lot of congestion if you are soldering a lot of components.

Multimeter

A multimeter is a useful device that measures electricity, just like you would use a ruler to measure distance or a stopwatch to measure time. The best thing about a multimeter is that it also measures a lot of different things, such as voltage, current, resistance, and much more. A standard multimeter will have a large dial in the middle, which lets you select what you want to measure.

Most multimeters can measure voltage, current, and resistance; some multimeters also have a continuity check, which tests to see if the electrical circuit is complete by producing a loud beep when two things are electrically connected. This is helpful in diagnosing problems with a circuit—you can trace the voltage around the circuit and find which part is incomplete or not functioning the way it should. In contrast to this, you can make sure that two things are not connected just in case you don't want a certain part of your circuit to short, or you may want to test your soldering skills by not accidentally soldering joints together.

There are some advanced multimeters, which are often expensive, that have certain additional functions, such as the ability to measure transistors

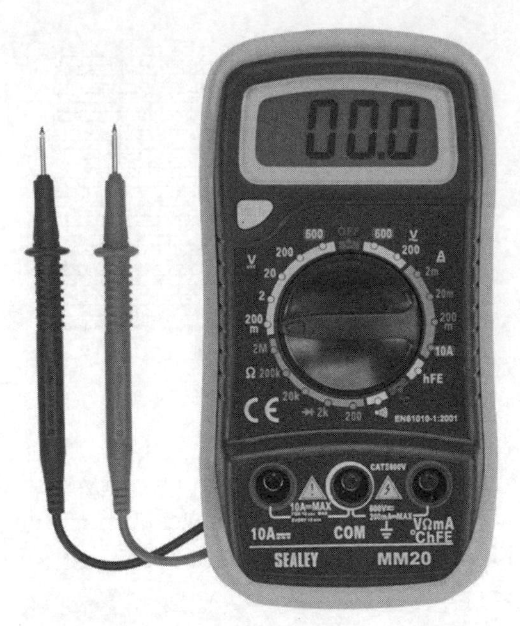

Figure A.4 *A typical multimeter.*

or capacitors. These functions are more suited for the professional engineers who design and manufacture high-end products.

In order for us show to you how a multimeter works, it is important to understand what we are measuring such as voltage, current, or resistance:

- Voltage is how hard the electricity is being pushed through a circuit—a higher voltage is being pushed through a circuit really hard. Voltage uses the symbol V.
- Current is how much electricity is flowing through a circuit—a high current indicates that more electricity is flowing through. The symbol for current is A.
- Resistance is how difficult it is for electricity to flow through a circuit—a higher resistance indicates that it is much more difficult for it to flow through. Resistance is measure in ohms, and the symbol for resistance is Ω.

It is also worth noting that the symbol that is being used for a unit may differ from a symbol for a variable equation.

Soldering

Learning how to solder is an essential skill in the world of electronics. Although you can get by just using a prototyping board, you may still need to solder headers onto the board or make some small modifications to a component.

Solder refers to the alloy that typically comes on a wire spool or tube, and it is this solder that we use to fuse components together on a printed circuit board. When selecting solder it usually comes in two types: leaded and lead free. When solder was first around, it was generally made up of an alloy using both lead and tin; since then it has become known that lead can be quite harmful when exposed to in large amounts. Lead was used in solder because it has a great low melting point and it created really good solder joints, which produced a highly reliable circuit board. Unfortunately, in the European Union leaded solder is not Restriction of Hazardous Substances (RoHS) compliant, and this restricts the use of leaded solder in electrical equipment and hence why lead-free solder is commonly used. Lead-free solder is usually made up of other metals such as silver and copper. Lead-free solder does come with its own downfalls—for example, it has a much higher melting point because of the tin content and as such requires a high-powered soldering iron.

Lead-free solder often contains a flux core, which helps give the same quality effect as leaded solder. Flux is a chemical agent that aids in flow and creates much better contacts when finished.

Figure A.5 *Solder spool.*

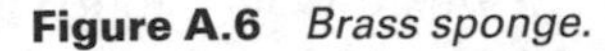

Figure A.6 *Brass sponge.*

Many tools are used to aid in soldering, but none are more important than a soldering iron. Soldering irons come in a variety of types, ranging from basic soldering irons to complex soldering stations, but they all serve the same function and purpose. Usually, a good place to start is to buy a station that comprises a soldering iron with either a digital or analog controller and a stand. These stations are becoming more common now and can be inexpensive to purchase at your local store.

Over time your soldering tip will start to oxidize and will turn black; this is bad because the soldering iron will not cling to the solder and you won't be able to solder a component; this is more commonly found with lead-free solder. This is where a simple soft sponge comes to the rescue—every so often you should clean the tip by wiping all the excess off. For even better results, you can use a brass wire sponge.

Apart from a soldering iron and solder, several other great accessories can aid in the process of soldering. A solder wick is a vital tool for mopping up if you have made a bit of a mess; you can also use it for desoldering. Solder wick is made up of thin copper braiding, and just like any PCB, it will soak up the solder, erasing any excess drops.

Figure A.7 *Tip tinner.*

You can also use a tip tinner to clean your tip. It is composed of a mild acid that helps remove any unwanted residue left on your soldering tip and prevents oxidization when the tip is not in use.

As previously mentioned, some lead-free solder comes with a core flux; however, sometimes it is not enough and extra flux may be required. Flux pens are used to allow difficult components to create a better bond to the PCB.

Analog versus Digital

Analog and digital signals are used to transmit an array of information, usually conveyed through electrical signals. The main difference between the two signals is that analog signals are transmitted in pulses of varying amplitude, and digital signals are transmitted into a binary format such as a one or zero, where each bit represents a distinct amplitude.

Analog refers to the circuits in which quantities such as voltage or current vary at a continuous rate over a period of time. Electronic signals represent information by changing their voltage or current over time. The signal takes any value in a given range, and each signal value can represent a different kind of information. Any change that takes place in the signal has a significant impact on the overall result.

Something important to take into account is that analog signals can create noise, which is classified as a disturbance or variation, which can be caused by thermal vibrations. Because any slight variation in the signal can affect the outcome, this noise can have a significant effect, especially over long distances as the signal degrades.

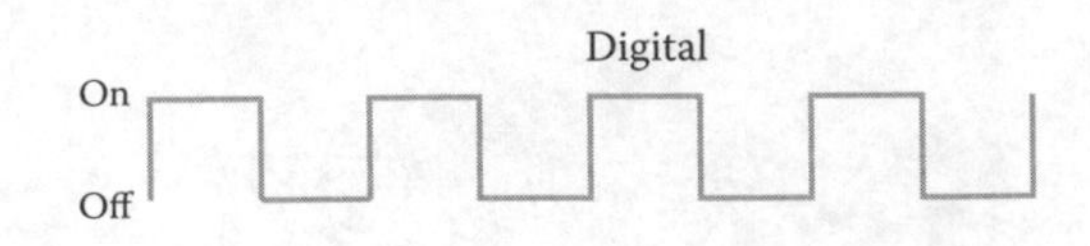

Figure A.8 *Digital signal.*

When designing a system, analog circuits are much harder and complex and require more skill compared with digital systems. This is primarily why digital systems have become more common; in addition, they are much cheaper to manufacture.

Digital systems are much easier to understand—they do not use a continuous range like analog, so any noise or slight variation in the signal does not affect the result of a digital signal. Digital systems generally have only two states, and they are represented by two different voltages—usually 0 is equal to ground and 1 is equal to +V.

The main advantage of using a digital system is that when compared with analog, the signal does not degrade over time and it can be quite easily replicated without any loss. The main disadvantage is that digital circuits consume much more power than do analog circuits, and when circuits consume more power, that generally means more heat, which in turn increases the complexity of designing a circuit.

All of the parts used in this book are easily available for purchase from a number of stores through the Internet. However, sometimes it can be quite difficult to find exactly what you are looking for, especially in your own country. Finding suppliers in your own country can reduce the cost of shipping, offering you a more affordable solution.

Suppliers

When searching for your components on the Internet, you will probably come across many suppliers, all offering something different in terms of product selection. Photon boards are available from a wide range of suppliers, as well as through Particle itself (www.particle.io.com); you will also find a range of kits and shields and other interesting Photon products.

Most of the components used in this book can be found in the Photon Maker Kit, which is available from most suppliers or directly from Particle. This is the easiest and most convenient way of sourcing the parts for this book. All of the components are also available from many other suppliers, such as Adafruit Industries, SparkFun, or SeeedStudios. These companies often specialize in the maker market and manufacture their own products.

SparkFun, based in the United States, is an online retail store that sells various pieces to make your electronic projects. In addition to their products, they provide classes and online tutorials designed to help educate individuals in embedded electronics (www.sparkfun.com).

Adafruit Industries was founded in 2005 by MIT engineer Limor Fried. Adafruit was designed to create the best place online for learning about electronics and making the best-designed products on the market. Adafruit designs and develops their own products in house; they sell these on their website, along with several tutorials to get you started (www.Adafruit.com).

SeeedStudio is based in Shenzhen, China, and the United States. Having access to the Far East market makes them a prime supplier for all hardware components. Not only does Seeed have a wide range of electrical components at very low cost; they also offer a wide range of services, from manufacturing to 3D printing to laser cutting and more. Check out their website for further information (www.seeedstudio.com).

The sections that follow list components by type along with some sources and order codes to make it easier for you to purchase your components.

Components

The tables for each project list appendix codes for each component that is used. This section lists all the parts and offers some sources as to where they can be purchased.

Resistors

Resistors are low-cost components—almost less than one cent each; often suppliers will sell them in packs of 50 or 100. There are common resistors that

Code	Description	Source
M1	Photon board	Particle SeeedStudio: 114990286 SparkFun
M2	Ultrasonic sensor	SeeedStudio: 101990004
M3	Particle relay shield	Particle
M4	Photon Grove Kit	SeeedStudio: 110060123

Table A.1 *Photon Kits and Modules*

Code	Description	Source
R1	220-Ω ¼-W resistor	Particle Kit SeeedStudio SparkFun Adafruit
R2	10-kΩ potentiometer	Particle Kit SeeedStudio SparkFun Adafruit
R3	10-kΩ ¼-W resistor	Particle Kit SeeedStudio SparkFun Adafruit
R4	Photocell	Particle Kit SeeedStudio SparkFun Adafruit
R5	4.7-K ¼-W resistor	SeeedStudio SparkFun Adafruit

Table A.2 *Resistors*

get used a lot, such as 220R, 270R, 1-K, and 10-K values, so it can be useful to keep several of these values on hand.

After a while you might find yourself buying a lot of resistors, and in some cases it is better to buy them in kits, which stock the most popular values used in everyday electronics.

Some companies that sell resistor kits are

- SeeedStudio: 110990043
- SparkFun: COM-10969

Semiconductors

The projects in this book use a lot of light-emitting diodes (LEDs), so sometimes it is worth looking around for a variety pack of 5-mm or 10-mm LEDs rather than buying all the size and color combinations separately.

Hardware and Miscellaneous

Most hardware and in particular some miscellaneous parts can be found in most maker/hobbyists stores worldwide.

Code	Description	Sources
S1	5-mm LED	Particle Kit SeeedStudio SparkFun Adafruit
S2	Temperature sensor DS18B20	SeeedStudio SparkFun Adafruit
S3	10-mm RGB LED	SeeedStudio SparkFun Adafruit

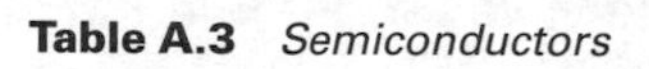
Table A.3 *Semiconductors*

Code	Description	Sources
H1	Breadboard	Particle Kit SeeedStudio: 319030001 SparkFun: PRT-12002 Adafruit: 64
H2	Jumper wires (male/female)	Particle Kit SeeedStudio:110990029 SparkFun Adafruit: 153
H3	Digital multimeter	SeeedStudio:405010002 SparkFun: TOL-12966 Adafruit: 308
H4	16 × 2 character LCD	SeeedStudio: 104990004 SparkFun: LCD-00790 Adafruit: 181
H5	5-V servo motor	Particle Kit SeeedStudio: 108090000 SparkFun: ROB-10333 Adafruit: 155
H6	Tactile push switch	Particle Kit SeeedStudio OPL SparkFun: COM-10302 Adafruit: 1119
H7	9-V DC power supply	Electrical stores
H8	9-V lamp	
H9	9-V PP3 battery	Most electrical retailers
H10	Equipment wire	Most electrical retailers

Table A.4 *Hardware and Miscellaneous*

B

Particle Code Reference

This appendix will give you a basic understanding of different functions within the Particle code, their correct syntax, return values, and a basic example of how to use that function within your code.

Setup

This is a block of code that is executed only once your Photon board has booted up.

Example

```
void setup() {
//code is executed here only once
pinMode(D0, OUTPUT);
}
```

Loop

This is part of the code where it is executed repeatedly after the initial setup function.

Example

```
void loop () {
      digitalWrite(led, HIGH);
      delay(1000):
      digitalWrite(led, LOW);
      delay(1000);
}
```

This continuously turns a light-emitting diode (LED) on and off every 1 second.

analogRead

This obtains the value of a particular analog pin on the Photon board. The values range from 0 to 4095, where 4095 is 3V3. See Chapter 5 for more information on analog inputs.

Syntax

```
analogRead(pin);
```

Parameters

pin

The Photon board pin numbers A0–A5.

Return Value

An integer between 0 and 4095.

Example

```
int temperature = analogRead(pin);
```

This obtains the value from the analog pin and stores it in an integer variable called temperature.

analogWrite

This sets the duty cycle of an analog pin, which is capable of pulse width modulation (PWM). Remember we are using a digital system, so we can only emulate an analog signal using PWM. This function sets the pin value between 0 and 255, where 0 is GND and 255 is 3V3.

Syntax

```
analogWrite(pin, value);
```

Parameters

pin

The pin number.

value

An integer between 0 and 255.

Return Value

None

Example

```
analogWrite(A7, 127);
```

This pulses analog pin 7 so that it is turned off 50% of the time. If you are using an LED, this will be half the brightness.

digitalRead

This reads the value of a digital input and returns a value of HIGH or LOW, meaning the pin is either ON or OFF.

Syntax

```
digitalRead(pin);
```

Return Value

HIGH or LOW

Parameters

pin

The digital pin number.

Example

```
if (digitalRead(D0) == HIGH) {
    Serial.println("Pin 0 is HIGH")
}
```

This prints out in the serial box "Pin 0 is HIGH" when digital pin 0 is connected to 3V3.

digitalWrite

This function sets the digital pin to either HIGH or LOW, where LOW is GND and HIGH is 3V3.

Syntax

```
digitalWrite(pin, value);
```

Parameters

pin

The digital pin number.

value

The value is either HIGH or LOW.

Return Value

None

Example

```
digitalWrite(D0, HIGH);
```

This turns on digital pin 0 and outputs 3V3.

if

This very useful function executes a certain block of code if the conditions are true.

Syntax

```
if(condition) {
      //executable code goes here
}
```

Example

```
int = 1;
if (n < 1) {
      digitalWrite(Ledred, HIGH);
}
if (n > 1) {
      digitalWrite(Ledgreen, HIGH);
}
if (n == 1) {
      digitalWrite(Ledyellow, HIGH);
}
```

else

This is used in conjunction with the `if` statement. The block of code within the `else` statement will execute when the `if` statement returns a false condition or when the `if` condition is not met.

Syntax

```
if (condition) {
//execute if condition is true
}
else {
//execute if condition is false
}
```

Example

```
int switchstate = digitalRead(D0)
     if (switchstate == HIGH) {
          digitalWrite(ledpin, HIGH)
     }
     else {
          digitalWrite(ledpin, LOW)
     }
```

This example is designed to detect if a switch has been pressed. If the switch has been pressed and equals HIGH, then the LED turns on. If the switch is equal to LOW, then the LED will stay off.

int

An integer is a data type that creates a bit of memory to store a single value.

Example

```
int led = D0;
```

This stores the pin number into a memory slot called led. When you want to reference the pin number, instead of typing the pin number every time, it is easier to put "led" if you have an LED connected to that pin.

pinMode

This sets the direction of a pin to either an output or an input. This function is always included in `setup()`.

Syntax

```
pinMode(pin, mode);
```

Parameters

pin
  The pin number.

mode
  Either an INPUT or an OUTPUT (case sensitive).

Example

```
pinMode(D0, OUTPUT);
```

This sets digital pin 0 to an output pin, which can be controlled using the function `digitalWrite`.

servo.attach

This assigns a servo object to a particular pin on the Photon board.

Syntax

```
myServo.attach(pin);
```

Parameters

pin

The pin number.

myServo

Represents any servo object.

Example

```
Servo myServo;
void setup () {
      myservo.attach(D0);
}
```

servo.write

This sets the exact position of a servo motor.

Syntax

```
myServo.write(angle);
```

Parameters

angle

The angle to set the servo to—usually 0 to 180 or –180 to 180.

myServo

This represents any servo object connected.

Example

```
Servo myServo;
void setup() {
      myServo.attach(100);
}
```

INDEX

Numbers

A

D

W